Venus
(and How to Terraform it)

Venus
(and How to Terraform it)

By

'The Science Geek'
(former name of Steve Hurley's Explaining Science blog

VENUS

First published as a kindle e-book December 2018
Revised edition May 2019
Minor revisions Oct 2019

© The Science Geek 2018, 2019
All rights reserved

ISBN: 9781790904655

Dedication

I dedicate this book to my wife Julia Marwood. She was responsible for editing the manuscript and suggesting numerous improvements and corrections. Without her there would have been no book on terraforming Venus.

Contents

Introduction

Chapter 1 Venus - the Morning Star and the Evening Star 1

Chapter 2 Exploring Venus..15

Chapter 3 Transit of Venus..27

Chapter 4 Living on Venus 41

Chapter 5 Terraforming Venus 49

Chapter 6 Planetary Magnetic Fields 61

Chapter 7 Giving Venus a Magnetic Field 67

Chapter 8 Conclusions 73

Appendix I: Glossary..75

Appendix II: Determining the size of the AU..81

Appendix III: Venus Facts 83

References 91

About the author 93

Introduction

The planet Venus is our nearest neighbour, the planet closest to the Earth in size and internal composition, and the brightest natural object in the sky after the Sun and the Moon. Often called Earth's sister planet, Venus is a mysterious object whose secrets were not unlocked until the 1960s.

The first chapter focuses on how Venus appears from Earth. The 584-day cycle where Venus goes through a full set of phases like the Moon is explained in detail. The fact that Venus's phases can only be explained by a heliocentric theory and not the geocentric theory, something which was very important to the development of astronomy, is discussed.

I then talk about the early depiction of Venus in 1950's science fiction as a world on which exotic life forms could survive. However, exploration of Venus by spacecraft, discovered a harsh hostile world, where the surface temperature is nearly five hundred degrees Celsius and the atmospheric pressure is a crushing 92 times that of Earth. Despite this I will argue later in the book that it is possible to terraform the planet to make it more Earth-like and may be desirable to do so.

I discuss the 'transit of Venus', an astronomical event in which Venus appears to cross the surface of the Sun. Observations of the transit of Venus were critical to the development of astronomy in the seventeenth and eighteenth centuries.

The remainder of the book discusses what humans would need to do to live on Venus. The idea of 'floating cities' is discussed and terraforming is explored in detail and how the challenges posed by such a hostile environment could be overcome.

A glossary of terms is included at the end of the book, together with additional information about the planet sourced from NASA.

Steve Hurley

Venus and How to Terraform it

www.explainingscience.org

October 2019

Venus and How to Terraform it

Chapter 1

Venus - the Morning Star and the Evening Star

At certain times anyone looking at the eastern sky before sunrise will notice a brilliant white object – the planet Venus, sometimes called the Morning Star. It is much brighter than any other star or planet and is the third brightest object in the sky after the Sun and the Moon. On other occasions the same brilliant white object will appear in the western sky after sunset, when it is known as the Evening Star.

This image from NASA shows Venus as the 'Morning Star' in the pre-dawn sky. *The less bright object near to Venus is the planet Jupiter, which is about 10 times fainter. Venus appears as a disc, rather than a point of light, in the image because it is so bright that it has flooded the camera with light. To the naked eye observer Venus is so small that it appears as a point of light.*

There are three reasons why Venus appears so bright. Firstly, it is relatively large at 12,100 km in diameter, which is 95% of the diameter of the Earth and much larger than the two other inner rocky planets, Mars and

Mercury. Secondly, it is surrounded by thick clouds, which reflect 76% of the sunlight hitting the planet back into space.

Finally, it is the planet which passes closest to Earth. At its nearest approach it is only 40 million km from Earth, roughly 100 times the distance from the Earth to the Moon (Williams 2017).

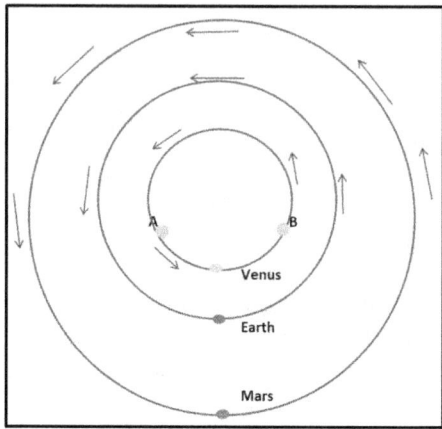

The orbit of Venus is shown in the diagram above. It is closer to the Sun than the Earth and moves faster in its orbit. A year on Venus lasts only 225 Earth days. Because Venus orbits inside the Earth's orbit, to an observer on Earth, it always appears close to the Sun. This means that for most of the time Venus is only above the horizon in the daytime when it is difficult to see. The points marked as A and B in the diagram are known as the greatest elongations and are the times where Venus appears to be at its greatest distance from the Sun in the sky. At these times, although Venus is still visible mainly in the daytime, it is also visible for a few hours at night before sunrise (point B) or after sunset (point A).

Mercury also orbits the Sun inside the Earth's orbit and like Venus is only visible for a short time before sunrise or after sunset. The other five planets (Mars, Jupiter, Saturn, Uranus and Neptune) lie outside the Earth's orbit and therefore can be seen in the middle of the night.

How Venus moves as seen from Earth

As mentioned above, Venus takes 225 days to complete one orbit around the Sun. The Earth, which is further out than Venus, and orbiting at a slower speed, takes just over 365 days. It takes Venus on average 584 days to return to the same position with respect to the Earth and the Sun. Astronomers call this interval of time its **synodic period**. The diagram below shows how, as seen from Earth, the visibility of Venus changes during this 584-day cycle.

- If we take the starting point **A** in the diagram, then this date is one of the two greatest elongation points, where Venus appears furthest from the Sun in the sky. At this time, Venus appears as the **Morning Star** clearly visible before sunrise as a brilliant white object in the eastern sky. After the Sun has risen Venus becomes difficult to see in the bright daytime sky. However, it doesn't disappear altogether. If you know exactly where to look, Venus can be seen in the daytime as a faint white dot.

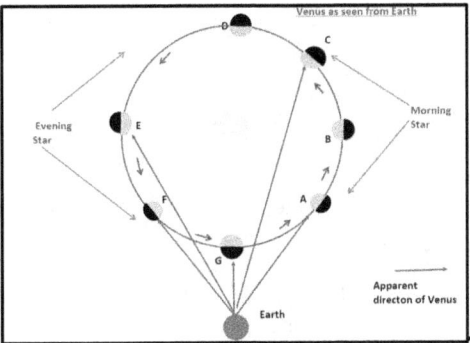

How the position of Venus changes over the 584-day cycle.

Table of rising and setting times of the Sun and Venus for London at points A to G during 2017 and 2018

	Date	Visibility	Sun Rise	Sun Set	Venus Rise	Venus Set
A	3 June 2017	Morning Star (Greatest elongation)	4:46 AM	9:10 PM	3:10 AM	4:42 PM
B	10 August 2017	Morning Star	5:37 AM	8:32 PM	2:26 AM	6:38 PM
C	30 November 2017	Morning Star	7:41 AM	3:55 PM	6:47 AM	3:29 PM
D	9 January 2018	Visible daytime only (Superior conjunction)	8:02 AM	4:00 PM	8:10 AM	4:04 PM
E	10 June 2018	Evening Star	4:43 AM	9:16 PM	7:24 AM	11:51 PM
F	17 August 2018	Evening Star (Greatest elongation)	5:48 PM	8:18 PM	10:16 AM	9:28 PM
G	26 October 2018	Visible daytime only Inferior Conjunction	7:42 AM	5:44 PM	8:03 AM	4:56 PM

Data from http://www.mso.anu.edu.au/rsaa/observing/webephem/

- As Venus continues in its orbit, over the following months it appears to get gradually closer to the Sun and is visible for a shorter time before the Sun rises. By Point **C**, which occurs five months after point **A**, in London, it is only visible for about 50 minutes before sunrise.
- Five weeks later Venus has moved directly behind the Sun. This is called superior conjunction (point **D**) and, for a few weeks either side of this date, Venus is difficult to see because it is only visible in the daytime close to the Sun.
- As it continues in its orbit, it rises and sets later and becomes clearly visible as the bright **Evening Star** in the West (point **E**).

- Seven months after superior conjunction it reaches the other greatest elongation point, (point **F**) where it is a brilliant object in the early evening western sky
- After the greatest elongation, its apparent distance from the Sun starts to get smaller again and Venus is visible for a shorter time after sunset. Two months later it passes between the Earth and the Sun; this is called inferior conjunction (point **G**). For a few weeks either side of this date Venus is very difficult to see because it is only visible in the daytime close to the Sun.
- After inferior conjunction, Venus continue to rise earlier and is visible again in the morning as the **Morning Star** in the East, returning to point A where the cycle begins again.

Transit of Venus

Because the orbit of Venus is tilted with respect to the Earth's orbit, at inferior conjunction Venus normally passes just below or just above the Sun. Occasionally things line up so that Venus passes directly in front of the Sun as seen from the Earth. This is called a transit of Venus and the planet appears as a small dark dot against the bright disc of the Sun. This is a rare occurrence; it last happened in 2012, and it will not happen again until the year 2117 (McClure 2012). I will discuss transits of Venus in more detail in Chapter 3.

The phases of Venus

As seen from the Earth, Venus goes through a full set of phases in a similar way to the Moon although, because Venus appears so small, they are only visible through a telescope.

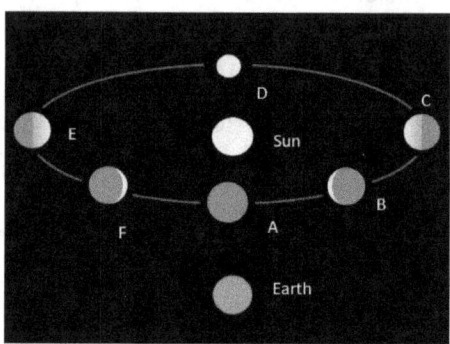

At inferior conjunction (A), Venus is between the Earth and the Sun and the sunlit part of Venus faces away from us making the planet difficult to see. This can be considered a New Venus in an analogous way to a New Moon. Over the cycle described previously, the sunlit part of Venus gets larger or *waxes* through to a crescent phase (B), to a Half Venus (C) at the greatest elongation. At superior conjunction (D), when the whole Earth-facing side is illuminated, it is in the Full Venus phase. Venus then *wanes* back to a Half Venus (E) at greatest elongation, then to a crescent (F) and finally back to a new Venus.

Unlike the Moon, which is always brightest at the Full Moon phase, Venus is not at its brightest at a Full Venus. The reason for this is that, at Full Venus, Venus is at its furthest distance from the Earth (around 260 million km) and is at its smallest in the sky. Similarly, when Venus is at its closest to the Earth the sunlit side faces away from us, so it is at its faintest. Venus is at its brightest 35 days before and after inferior conjunction, when it is a crescent with 26% of the planet illuminated and is at a distance of approximately 60 million km from Earth.

How bright does Venus get?

When discussing the brightness of objects in the sky, astronomers use a scale called magnitude, where the lower the magnitude the brighter the object. The scale was invented by the ancient Greek astronomers who classified all the stars visible to the naked eye into six magnitudes. The brightest stars were said to be of magnitude 1, whereas the faintest were of magnitude 6, which is the limit of human visual perception (without the aid of a telescope). Nowadays, of course, the situation is somewhat different – very large telescope such as the Keck telescope allow us to see very faint objects down to a magnitude of 26. Whereas the brightest objects had all been given the magnitude of 1, now there is differentiation between them, with some having values of less than one, right down to -26.7. The scale is also applied to all objects in the sky, including manmade satellites, planets, comets, asteroids, moons and galaxies, not just stars.

Values in the scale were defined by nineteenth century astronomers in order to make magnitude 1 a hundred times brighter than magnitude 6. Therefore, a *decrease in magnitude by 1* means an *increase in brightness by a factor of 2.512* and a decrease in magnitude by 2 would mean an increase in brightness by 6.31, because 2.512 x 2.512 = 6.31. So, for example:
- a star of magnitude 1 is **15.9** times brighter than a star of magnitude 4. This is because 2.512 x 2.512 x 2.512 = 15.9.
- a star of magnitude 1 is **100** times brighter than a star of magnitude 6. This is because 2.512 x 2.512 x 2.512 x 2.512 x 2.512 = 100

The brightest natural objects in the sky are (obviously) the Sun, which has a magnitude of -26.7, followed by the Moon, which has a magnitude of -12.7 at a typical full Moon. Third comes Venus, with a magnitude of around -4.5 at the two points in its orbit in its orbit when it is brightest. The fourth brightest natural object, and the second brightest planet, is Jupiter, which has a magnitude of -2.7 at its brightest. The brightest star in the sky is Sirius, but with a magnitude of -1.46, it is 16 times fainter than Venus.

Galileo's discovery

Nicolas Copernicus (1473-1543) published the theory of heliocentrism – that the planets orbit the Sun - in 1543, just before his death. He arrived at his conclusions through his own study of the sky with the naked eye and his interpretation of observations that had been made through the ages.

However, this theory was considered heretical by the Catholic church, which would not deviate from the theory of geocentrism, the view that the Earth was the centre of the Universe and that the stars, planets, Sun and Moon were in orbit around it. Certain verses in the Bible were put forward as evidence for this view, such as Psalm 104:5: 'the Lord set the Earth on its foundations; it can never be moved.'

It was the Italian astronomer Galileo Galilei (1564-1642) who, with the benefit of a telescope, discovered the phases of Venus in 1616. The way that it changes size throughout the different phases can only be fully explained by Venus orbiting the Sun, not the Earth. Galileo thus provided incontrovertible evidence that the geocentric theory was incorrect. Unfortunately for Galileo, the Catholic church was unconvinced, heliocentric books continued to be banned, and Galileo was ordered to refrain from holding, teaching or defending heliocentric ideas. Nevertheless, Galileo continued to defend heliocentrism, and in 1633 the Roman Catholic Inquisition found him 'vehemently suspect of heresy', sentencing him to indefinite imprisonment. He was kept under house arrest until his death in 1642.

However, the facts cannot be disputed. When viewed through a telescope Venus **does** show changes in size and shape which can only be satisfactorily explained by a heliocentric model. Eventually, in 1758, the Catholic Church dropped the general prohibition of books advocating heliocentrism.

Galileo - image from Wikimedia Commons

The properties of Venus

Venus appears to be a featureless object when seen through even the most powerful telescope. This contrasts with Mars, on which surface details can be seen with even a small telescope.

Venus through a telescope - Image from NASA

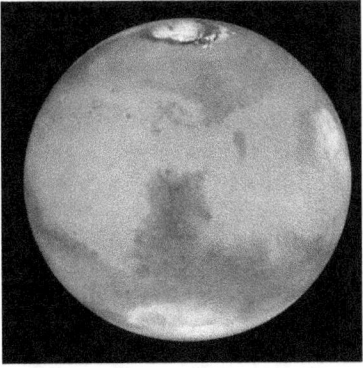

Mars through a telescope - Image from NASA

The reason why we cannot see any details on Venus is because the planet is completely covered in thick clouds which reflect most of the sunlight back into space and prevent us from seeing anything underneath. So, despite it being the closest astronomical object to the Earth other than the Moon, little was known about Venus until quite recently. Without any surface markings to follow as the planet spins it was not possible to determine how long a day on Venus was. Estimates ranged from around a day to several months. The true value wasn't discovered until 1961 when astronomers bounced radar signals off the planet and studied the echoes. The results were quite surprising -Venus rotates on its axis incredibly slowly, once every 243 days. Unlike most other planets it rotates in the opposite direction to which it orbits the Sun.

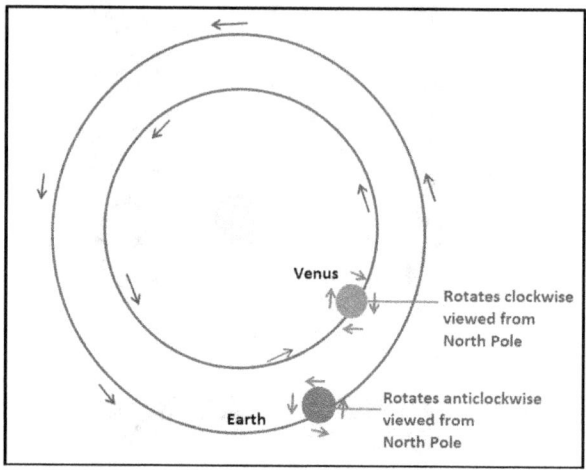

As seen from the North Pole, the Earth orbits the Sun in an anticlockwise direction and spins on its axis in an anticlockwise direction, whereas Venus orbits the Sun in an anticlockwise direction but spins on its axis in a clockwise direction.

Even after the rotation period was established, conditions below the clouds were unknown. We had to wait the planet was visited by spacecraft to find out how hot the surface and how thick the atmosphere were. In the next chapter, I will talk about these missions to Venus which unlocked its secrets.

How to see Venus during daytime

Although many people think that Venus can only be seen at night. It is perfectly possible to view it with the naked eye during the daylight hours, although it is difficult to see because the daytime sky is bright. When the Sun is above the horizon Venus appears as nothing more than a small white dot. One trick which you can use to find Venus during the daytime is to use the Moon as a reference point, at a time when the Moon is close to Venus in the sky. This is an event known as a conjunction of the Moon and Venus. The dates of conjunctions are given in astronomical tables which are freely available on the internet.

```
Apr 01  00:14   Moon at Apogee: 405577 km
    02  04:18   Venus 2.7°N of Moon
    02  23:01   Mercury 3.6°N of Moon
    05  08:50   NEW MOON
    09  06:40   Mars 4.7°N of Moon
    09  15:43   Aldebaran 2.1°S of Moon
    11  19      Mercury at Greatest Elong: 27.7°W
    12  18:08   Moon at Ascending Node
    12  19:06   FIRST QUARTER MOON
    13  20:12   Beehive 0.2°N of Moon
    15  00:24   Mars 6.4°N of Aldebaran
    15  08:22   Regulus 2.7°S of Moon
    16  20      Mercury 4.3° of Venus
    16  22:02   Moon at Perigee: 364209 km
    18  03      Venus at Aphelion
    19  11:12   FULL MOON
    23  00      Lyrid Meteor Shower
    23  00      Uranus in Conjunction with Sun
```

Part of a table of astronomical events for the year 2019 from http://www.astropixels.com/ephemeris/astrocal/astrocal2019gmt.html.

If we look at the table above, we see that there was a conjunction of Venus and the Moon on 2 April 2019. On this date Venus was 2.7 degrees north of the Moon; a distance roughly five times the Moon's diameter.

The next thing you need to do is to look at the position of Venus while it is still dark to get a feel of the relative positions of the Moon and Venus. On 2 April 2019 the Sun rose at 6:41 am local time in Manchester, England. So, if you had looked at the crescent Moon at about 6:00 am on this date, you would have seen Venus directly above it in the eastern sky and the sky would still be dark, making Venus easy to spot.

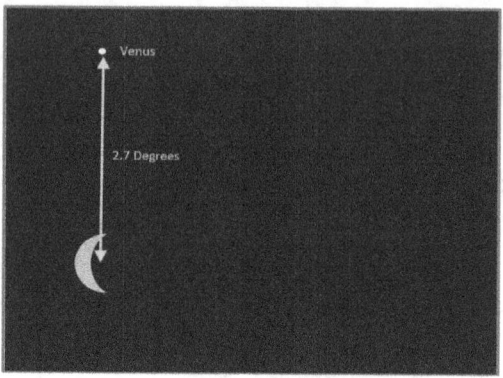

If you had looked at the Moon hours later when the Sun had risen, it would have moved across the sky in a westward direction. Venus would have appeared as a faint white dot in the same position relative to the Moon as earlier.

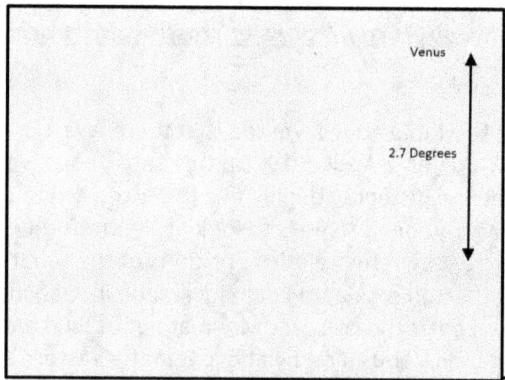

Chapter 1 - Notes

The 584-day cycle

If we consider Venus and the Earth in orbit around the Sun, then the synodic period is the interval of time it takes for Venus to do exactly one 1 extra orbit than the Earth. If we give the synodic period the symbol **p** then when the Earth has completed **p** orbits, Venus will have completed **p+1** orbits.

If we denote the amount of time that Earth takes to complete one orbit as O_E and the amount of time that Venus takes to complete one orbit as **Ov** then.

$$pO_E = (p+1) Ov \quad \text{(equation 1)}$$

If we multiply out the bracket and subtract **pOv** from both sides of equation 1, then we have:

$$p(O_E - Ov) = Ov \quad \text{(equation 2)}$$

Therefore

$$p = Ov / (O_E - Ov) \quad \text{(equation 3)}$$

If we put the actual value for **Ov** and O_E into equation 3, which are 365.256 days and 224.701 days respectively, then p is equal to 1.59867 Earth orbits or 583.92 days.

Chapter 2

Exploring Venus

Until the early 1960s the general view of Venus was that it was a planet with similar conditions to the Earth, but a little hotter because it is closer to the Sun. As mentioned in the previous chapter, Venus is surrounded by thick clouds and what was beneath the clouds was a complete mystery. This allowed writers to have a free rein in imagining conditions on its surface. In the years before the space age Venus had been depicted in fiction as being covered by deserts, swamps, oceans, jungle and rugged mountains, and all sorts of strange and exotic life forms were proposed to exist there.

As late as 1954, the American science fiction and popular science writer Isaac Asimov (1920-1992) wrote the novel *'Lucky Starr and the Oceans of Venus'*. In this he assumed that Venus had a temperate climate, a day length of 36 hours, an atmosphere that is 90% nitrogen and 10% carbon dioxide and a planet-wide ocean covering its surface, which was teeming with exotic life forms. The novel was set in the near future when humans had built colonies of millions of people living in huge domed cities on the Venusian ocean floor. Only a few years later, a very different picture emerged, and it was clear that Asimov's domed cities would not be happening any time soon.

Exploration of Venus

More accurate knowledge about Venus began to form when Mariner 2 became the first spacecraft to escape from the Earth's gravity and pass close to another planet. Launched on 27 August 1962, the spacecraft passed within about 34,000 kilometres (21,000 miles) of Venus on 14 December 1962.

Mariner 2 - image from NASA

    The instruments on Mariner 2 returned a lot of data about the planet. Readings from them showed that the atmosphere was much thicker than the Earth's and that the temperature of both the day and night side of the planet were around 240 degrees Celsius (Williams 2017b). This came as a great surprise and was also much hotter than astronomers had expected – in fact, the true temperature turned out to be even higher, and will be discussed in more detail below. Mariner 2 also made the unexpected discovery that Venus, unlike the Earth, has almost no magnetic field to shield the planet from the solar wind, the stream of electrically charged particles from the Sun. In a surprising omission Mariner 2 did not have a camera, so was unable to take any pictures of Venus. Even the smallest camera equipment would have added precious ounces to the weight of the spacecraft, and the decision not to send a camera was based on the reasoning that there would be little to see, and more significant data could be acquired with other instruments. Taking the camera would have required the removal of one or more of those instruments.

    During the 1960s and 1970s many more spacecraft from the Soviet Union and the USA flew past the planet and it became clear that the conditions were even harsher than Mariner 2 had suggested. On 15 December 1970 the Soviet spacecraft Venera 7 landed and transmitted 23 minutes' worth of data before being destroyed by the intense heat. On 22 October 1975 Venera 9 landed and became the first spacecraft ever to take pictures from the surface of another planet.

In 1989 the American spacecraft Magellan orbited Venus for four years and, using sensitive radar which was able to pass through the clouds, made a detailed map of about 98% of the planet's surface (Davis 2018). Magellan found that volcanic surface features, such as vast lava plains, fields of small lava domes, and large flat volcanoes (known as shield volcanos) are common. Interestingly, Magellan found few impact craters on Venus, suggesting that the surface is geologically young, only around 250 million years old.

A radar view of Venus from Magellan - image from NASA

Our current knowledge of Venus

Since the Mariner 2 mission, our knowledge of Venus has been transformed. We now know that the planet's atmospheric pressure is a crushing 92 times that of the Earth and consists of 96.5% carbon dioxide (Williams 2017a). The remainder of the atmosphere is mostly nitrogen. There is virtually no oxygen and only a small trace of water vapour. Venus rotates so slowly that a night on Venus lasts 58 Earth days. However, the thick atmosphere forms a blanket around the planet and means that the day and night temperature are the same, around 460 degrees Celsius – significantly hotter than the earlier readings had indicated, and hot enough to melt lead.

Interestingly, as its thick atmosphere smooths out temperature variation across the planet, there is no change with latitude. Thus, temperatures are the same at the equator and the poles of Venus. However, as on the Earth, the temperature and atmospheric pressure fall with altitude. The highest point on Venus's surface is called Maxwell Montes and is at an altitude of around 11km above the mean level of the planet's surface, about 30% higher than Mount Everest. At this altitude the temperature is a relatively mild 380 degrees.

At an altitude of about 50 km both the atmospheric temperature and pressure are like the values on Earth at sea level. So, it has been suggested by some science writers that in the future we could explore Venus by balloons which would float above the planet at this height. It has even been proposed that we might build colonies floating in the Venusian atmosphere. I will discuss these ideas more in Chapter 4.

Although no active volcanoes have been observed, most scientists believe that Venus is volcanically active. In 2014 the European Space Agency's Venus Express spacecraft observed temperatures spiking by several hundred degrees in some spots on the planet's surface. These hotspots ranged in size from one square kilometre to 200 square kilometres. The most likely cause of these is lava running across the surface of the planet (Howell 2015).

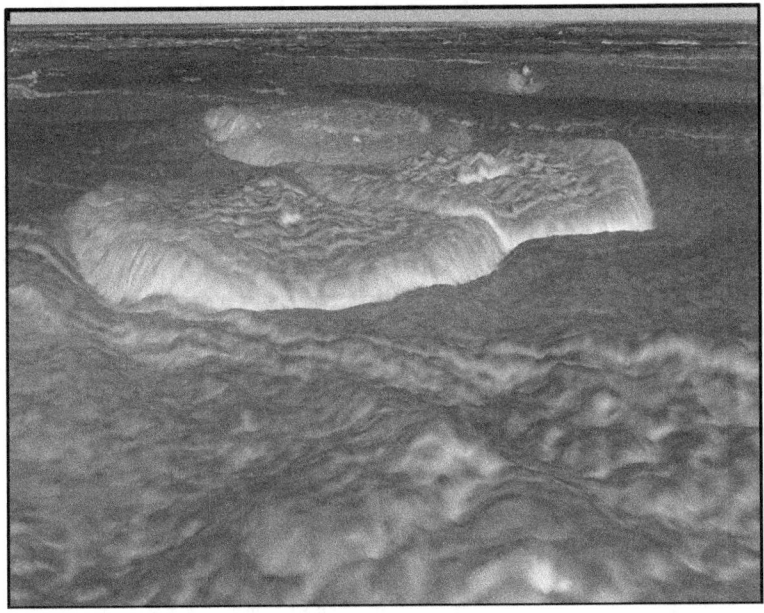

Computer generated image of lava flows on Venus - image from NASA.
Note the orange sky.

What would it be like to stand on the surface of Venus?

Assuming we were able to survive the high temperature and crushing pressure, we would see a landscape strewn with small rocks. Although the thick atmosphere would be relatively clear to look through, more distant objects would appear much more blurred and hazy than they would on Earth – in fact we would be able to see no more than a few kilometres. Despite being closer to the Sun than the Earth, the light levels would be relatively low - about the same as in a city like London or Manchester on a heavily clouded day. This is because the thick clouds made of droplets of sulphuric acid block most of the light from hitting the surface. Although an observer would be able to clearly tell the difference between night and day, they would never see the Sun, only the deep orange sky overhead

Why is Venus so hot?

The reason why Venus is so hot is because there is a huge amount of carbon dioxide in its atmosphere. This carbon dioxide acts as a powerful *greenhouse gas*. It lets sunlight pass through, but it reduces the amount of heat radiation which escapes back into space. It is called a greenhouse gas because a greenhouse gets very hot inside on a sunny day for the same reason. Light can pass through the glass, warming the objects inside, but the infrared rays emitted by these objects cannot pass back out.

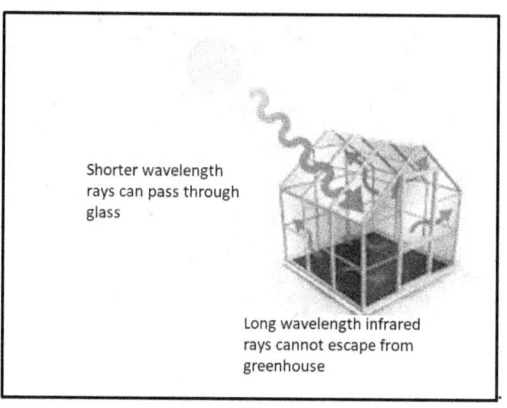

On Earth, carbon dioxide is only present in small amounts and its increase from 0.029% of the atmosphere, in the pre-industrial age, to its current value of 0.041% is generally accepted to be responsible for a rise in average global temperature of a few degrees. On Venus carbon dioxide makes up 96.5% of the atmosphere, and the atmosphere itself is much thicker than on the Earth, meaning that the greenhouse effect is enormous. It raises its surface temperature by an incredible 500 degrees above what it would be if it had an atmosphere with no greenhouse gases.

Some astronomers believe that billions of years ago, in the early days of the solar system, Venus and Earth had very similar conditions. Venus had plenty of water and may have had a global ocean. Because Venus is closer to the Sun it had always been hotter than the Earth. Although the Sun's energy output is often considered to be stable, this is not quite true - it is fact gradually increasing.

Although the present-day atmosphere of Venus is 97% carbon dioxide, this may not always have been the case. It may be that billions of years ago the increased energy from the Sun caused a warming of Venus's surface. Because water evaporates more rapidly at higher temperatures, there would have been an increased rate of evaporation from Venus's seas and oceans. This would have led to an increased concentration of water vapour in the atmosphere. Water vapour is a very efficient greenhouse gas and would have acted to trap heat escaping from the surface, making it even warmer. The temperature rise would have led to a higher surface temperature, causing water to evaporate at a higher rate.

The process is called a runaway moist greenhouse and is illustrated in the diagram below.

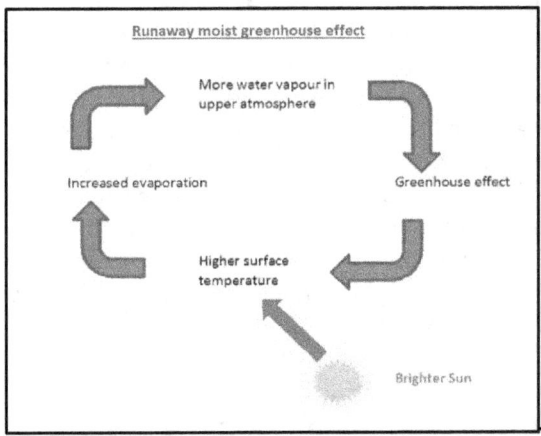

When it reached the upper atmosphere the water vapour would have been broken down by ultraviolet light from the sun into its component elements hydrogen and oxygen. Most of the hydrogen and oxygen produced would have recombined back into water vapour, but a small amount of the hydrogen in the upper atmosphere would have been blown away by the solar wind and escaped into space. It would have been the gradual removal of hydrogen by the solar wind over billions of years which caused all the water to be lost from Venus. The higher temperatures would have released carbon dioxide into the atmosphere, by breaking down rocks made from carbonates, thus maintaining and enhancing the greenhouse effect, even when all the water had gone.

Recent Missions to Venus

In the 1960s and 1970s there was a great deal of interest in exploring Venus and a total of 30 spacecraft were launched to the planet. Only about half of these missions were successful. Recently, the appetite for exploring Venus has waned compared to other places in the Solar System. Since the end of the Magellan mission in 1994, only two spacecraft have been successfully sent on missions to Venus: a probe called Venus Express launched by the European Space Agency in 2005 (which discovered the possible presence of volcanic activity) and a Japanese spacecraft called Akatsuki in 2010 - although three spacecraft have flown past Venus during this time on missions to other planets. By comparison, in the same period of time, a total of twenty-five spacecraft have been launched to Mars, which most astronomers believe is a more interesting object to explore, especially since the prospect of finding primitive forms of life has not been entirely ruled out. There are also the considerations that any spacecraft which lands on the surface of Venus will not be able to survive for long without being destroyed by the harsh conditions,
whereas spacecraft have landed and transmitted data from the surface of Mars for many years. Another limitation is that the thick clouds over Venus prevent orbiting spacecraft making any observations of the surface in visible light. All that can be seen is the top of the cloud layers.

Akatsuki - recovery from failure

The spacecraft Akatsuki - named after the Japanese word for dawn – is interesting, because it is the only time that a spacecraft has been recovered from what seemed like total mission failure and achieved most of its objectives.

Launch of Akatsuki in May 2010 – image from Wikimedia Commons

On 7 December 2010, Akatsuki reached Venus after a seven-month journey. Upon arrival, its main engine should have fired to slow the spacecraft down and put it into orbit around the planet. However, this failed to happen properly, causing Akatsuki to shoot past the planet and go into orbit around the Sun.

Fortunately, all was not lost. Other than the faulty main engine, all parts of the spacecraft turned out to be fully functioning, and the Japanese Aerospace Exploration Agency (JAXA) had another opportunity to get Akatsuki into orbit when it passed close to Venus on 7 December 2015, which was exactly 5 years after the first attempt. When a mishap like this occurs, the space agency does not usually get a second chance.

JAXA managed to get the spacecraft into orbit by using four of the eight small rocket motors which were only intended to fire for short periods to finely tune the spacecraft's position. The thrust or force provided by rocket motors is normally measured in units called newtons, usually abbreviated to N. The main engine which failed to fire had a thrust of 500 N, whereas each of these small motors only generated 20 N thrust. They had to fire for nearly 21 minutes to slow down the spacecraft enough to allow it to go into orbit, something they were never designed to do.

It was fortunate that, unlike most spacecraft, the small rocket motors and the main engine of the spacecraft used the same fuel, a liquid called hydrazine, and even more fortunate that they shared the same fuel tank. If the spacecraft had not been designed like this then the rescue operation could not have worked, but to the delight of space exploration lovers all over the Earth, not just in Japan, all went well and **Akatsuki** went into orbit.

What will Akatsuki achieve?

After the spacecraft was finally placed into orbit the science phase of the mission began and is, at the time of writing, still underway. Akatsuki has a whole host of instruments which will return useful data. Some of these are listed below.

- A special camera to study lightning flashes, which it will use when on the night side of Venus when these are easy to spot.
- An instrument to study the structure of high-altitude clouds to enable us to understand more about Venus's weather.
- An ultraviolet camera to study the distribution of specific atmospheric gases such as sulphur dioxide in ultraviolet light.
- An infrared defector which will peer through Venus's atmosphere to see heat radiation emitted from Venus' surface rocks and will help researchers to spot active volcanoes, if they exist.

Some interesting findings have already emerged. In 2017 JAXA and a team of researchers from the Hokkaido University used images from the Akatsuki orbiter to track strong winds in the low and middle cloud region, which extends from 45 to 60 kilometres in altitude. The wind speed was highest near the equator at a speed of 360 km/h, and the researchers named this phenomenon the 'Venusian equatorial jet', as it appears somewhat similar to the jet stream in the Earth's upper atmosphere. Another interesting finding was that Akatsuki was unable to find any evidence of lightning, despite looking at the night side of Venus with sensitive cameras for over a year. This negative result was an interesting finding because previous studies by the Soviet Venera spacecraft and Venus Express had found burst of low frequency radio waves, which on Earth are normally associated with lightning bursts (Lorenz 2018).

Photo of Venus taken in infrared light by Akatsuki – Image from JAXA

Up to date details on the Akatsuki mission can be found on the following website http://akatsuki.isas.jaxa.jp/en/

Chapter 3

Transit of Venus

On 6 June 2012, a transit of Venus occurred. This is rare astronomical event, which has only happened eight times since the invention of the telescope (NASA 2012). In a transit of Venus, the planet passes directly in front of the Sun and appears as a large black dot on its surface - slowly moving from one side to the other in about 3 hours. This chapter is about the transit of Venus and why it has been so important to the development of the science of astronomy.

Transit of Venus - image from NASA

Why transits of Venus are so rare.

The Earth takes slightly longer than 365 days, 365.256 days to be precise, to complete one orbit of the Sun. Venus, which is both closer to Sun and moves faster in its orbit, takes 224.701 days to complete one orbit. As discussed in chapter 1, the point in time when Venus is closest to the Earth and lies directly between the Earth and the Sun is called inferior conjunction. The time interval between one inferior conjunction and the next is on average about 584 days. This is the time it takes for Venus to 'gain a lap' in its orbit around the Sun. This is illustrated in the diagrams below, which shows the relative motion of Venus and Earth in the years 2017 and 2018.

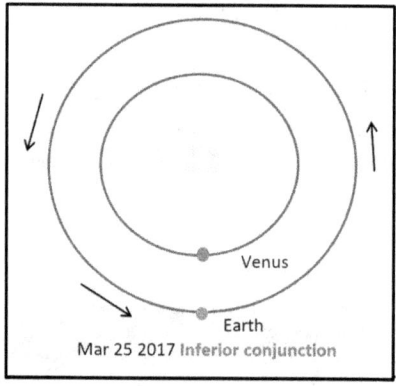

Venus and How to Terraform it

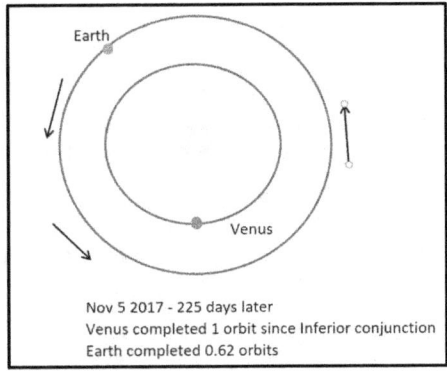

Nov 5 2017 - 225 days later
Venus completed 1 orbit since Inferior conjunction
Earth completed 0.62 orbits

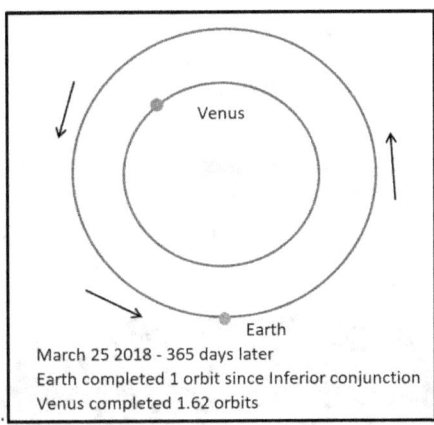

March 25 2018 - 365 days later
Earth completed 1 orbit since Inferior conjunction
Venus completed 1.62 orbits

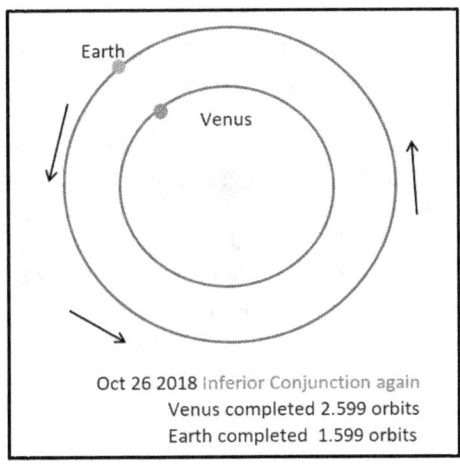

Oct 26 2018 Inferior Conjunction again
Venus completed 2.599 orbits
Earth completed 1.599 orbits

.Because the interval between inferior conjunctions is not an exact number of years, the date in the year on which an inferior conjunction falls varies from year to year. The dates of the inferior conjunctions between 2017 and 2023 are given below (Espenak 2014)

- 25 March 2017
- 26 October 2018
- 3 June 2020
- 9 January 2022
- 13 August 2023

See notes at the end of this chapter for more information.

If we look in three dimensions, then the orbits of Venus and the Earth do not lie in the same plane. Instead, the orbit of Venus around the Sun is tilted by a small angle, 3.4 degrees, with respect to the Earth's orbit.

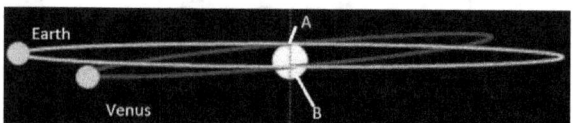

This means that, during the vast majority of inferior conjunctions, Venus does not pass directly between the Earth and the Sun. Instead, it will be just below or just above the Sun in the sky. There are two points in Venus's orbit when it intersects the plane of the Earth's orbit. These two points are known as the **nodes** and are marked as A and B in the diagram. An inferior conjunction can only give rise to a transit of Venus if it occurs within a small time-window a few days either side of the nodes. Point A is known as the descending node and occurs on 7 June each year. Point B is known as the ascending node and occurs on 9 December each year. The next time we will be able to see a transit of Venus will on be 11 December 2117, so I will not be around to see it.

Transits of Venus 2000 to 3000 CE
8 June 2004
6 June 2012
11 December 2117
8 December 2125
11 June 2247
9 June 2255
13 December 2360
10 December 2368
11 June 2490
10 June 2498
16 December 2603
13 December 2611
15 June 2733
13 June 2741
16 December 2846
14 December 2854
16 June 2976
14 June 2984

Data from NASA (2012). *Interestingly if you look at the dates in the table it shows another effect, which is that the time window around the nodes in which transits can occur drifts later by about 5 days every 500 years.*

Why do transits occur in pairs 8 years apart?

One numerical coincidence is that in when the Earth completes 8 orbits of the Sun – 2,922 days –Venus completes almost exactly 13 orbits (13.004 to be precise). Or, to put it another way, 8 Earth years is equal to almost exactly 13 Venus years. This means that every 8 Earth years, because almost (but not exactly) a whole number of Venus years have passed, an inferior conjunction occurs on the same date in the year minus 2/3 days. Therefore, as shown in the table above, transits of Venus generally occur in pairs eight years apart. For example, there was an inferior conjunction on 8 June 2004 which was also a transit of Venus. Eight years later the inferior conjunction on 6 June 2012 was also a transit of Venus. The inferior conjunction sixteen years later, on 3 June 2020, will not be a transit, as Venus will be just below the Sun.

Importance of the transit of Venus to astronomy

In the early 17th century the German astronomer and mathematician Johannes Kepler (1571-1630) worked out the relationship between the distance of a planet and the speed it orbits around the Sun, a relationship we now know as Kepler's laws.

Johannes Kepler - image from Wikimedia commons

Kepler's first law states that the planets move in ellipses with the Sun at one of the focuses of the ellipse. This is shown in the diagram below.

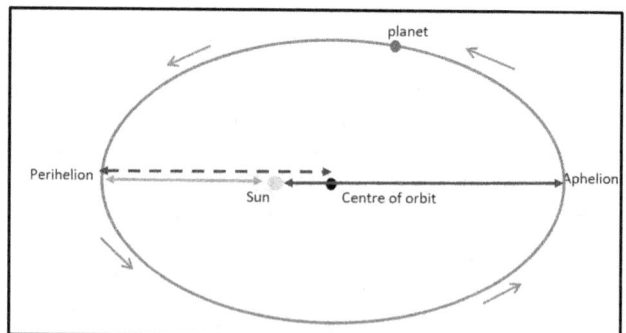

- The point where the planet is closest to the Sun is called its perihelion.

- The point where it is furthest away is its aphelion.
- Half the length of the long axis of the ellipse is known as the semi-major axis. It is equal to the average of the perihelion and the aphelion.

Kepler's second law states that an imaginary line drawn between the Sun and a planet covers equal areas at equal times. This statement is explained in the diagram below. As the planet moves around its orbit it moves:
- ***faster*** when it is ***closer*** to the Sun (when the line is shorter)
- ***slower*** when it is ***further*** away (when the line is longer).

The relationship between the planet's speed and its distance from the Sun is such that the area of the triangles shown in the diagram are the same.

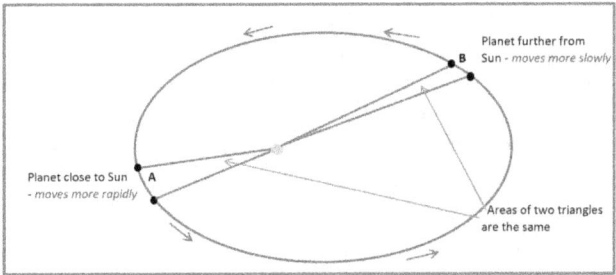

Kepler's third law states that the square of the orbital period of a planet is directly proportional to the cube of the semi-major axis of its orbit.

The third law is useful because it allows the **relative** distances to the planets to be calculated. For example, to work out the distance between Jupiter and the Sun relative to the Earth and the Sun, the following calculation can be used:

$$(P_J/P_E)^2 = (R_J/R_E)^3 \text{ (equation 1)}$$

which can be rearranged to give

$$R_J = \sqrt[3]{[R_E^3 (P_J/P_E)^2]} \quad \text{(equation 2)}$$

Where

- R_J is the average distance between the Sun and Jupiter
- R_E is average distance between the Earth and the Sun)
- and is equal to 1 astronomical unit (AU) - which is defined as the mean distance between the Earth and the Sun
- P_E is the time it takes the Earth to orbit the Sun - 1 year
- P_J is the time it takes Jupiter to orbit the Sun - 11.86 years

Putting the actual values into this equation gives us:

$$R_J = \sqrt[3]{[(1)^3 (11.86/1)^2]}$$

So R_J, the average distance between the Sun and Jupiter, is the cube root of 11.86 squared which is 5.20 AU. If we put the numbers in for some other planets, we get an average distance of 0.72 AU from the Sun fpr Venus and for Mars 1.52 AU. However, in Kepler's time it was not known, to any degree of accuracy how large the AU actually was. If astronomers could find out how big the AU was, they could work out the **real** size of the Solar System and the distances between all the objects within it. Determining the size of the AU became one of the key problems in 17th and 18th century astronomy.

Astronomers then realised that observations of a transit of Venus can be used to calculate the size of the AU. The calculations rely on the fact that Venus follows a slightly different path across the Sun when observed from different places on the Earth. If the start and end times of the transit are accurately timed in two different places and the distance between the places is known, then the distance between the Earth and the Sun can be calculated. More details are given in Appendix II

The first person to measure the size of the AU, based upon observations of a transit of Venus, was the British astronomer Jeremiah Horrocks (1618-1641), pictured below. He observed the 1639 transit and estimated a value of 96 million km by combining his observations made near Preston, in the North West of England, with those made at the same time by a colleague William Crabtree 60 km away in Broughton near Manchester.

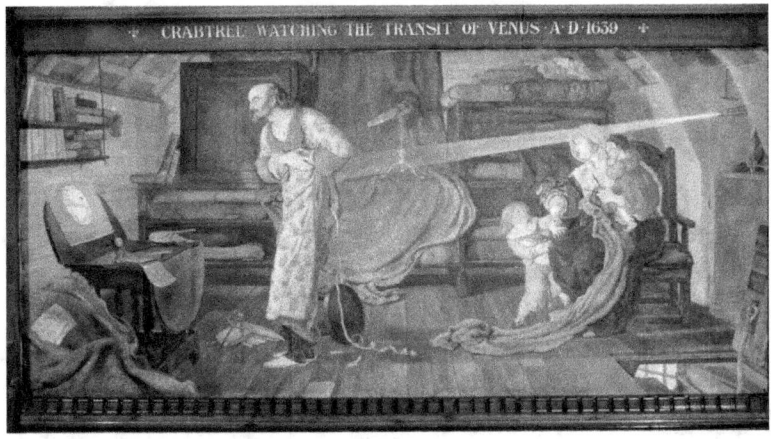

Mural in Manchester Town Hall of Horrock's colleague William Crabtree watching the 1639 Transit

Horrocks's value was about 36% lower than the correct value and was later refined by observations of the 1761 and 1769 transits to 94.3 million miles (151.7 million km) which is within 1.5% of the accepted value today.

Once the size of the AU was known, it was possible to determine the real size of the Solar System, and also the mass of the Sun. For this latter task, it was necessary to use Isaac Newton's laws of gravity, by which a relationship between three quantities can be derived.
- the period that a smaller object (such as the Earth) takes to orbits a larger object (such as the Sun)
- the distance between the two objects and
- the mass of the larger object.

This relationship is mathematically expressed as

$M = (4\pi^2 D^3)/(GP^2)$
where
- M is the mass of the larger object
- D is the distance between the two objects - in the case of the Earth and Sun, this is roughly 150 million kilometres

- P is the orbital period of the smaller object around the large one - in the case of the Earth around the Sun it is 365.25636 days or 31,558,150 seconds
- G is a number called Newton's gravitational constant and is equal to 6.674 x10^{-11} (in modern units of cubic metres per kilogramme per square second)

So once the real distance from the Earth to the Sun was known, by using this equation, the mass of the Sun could be worked out. It turned out to be 1.9885×10^{27} tonnes. This is roughly 330,000 times greater than the mass of the Earth.

A further discovery from the 1761 transit of Venus

The Russian astronomer Mikhail Lomonosov observed the 1761 transit and reported a bump or bulge of light off the solar disc as Venus began to exit the Sun. He correctly attributed this bulge to the bending of sunlight by an atmosphere around Venus. This was the first time that an atmosphere had been found around any planet other than the Earth.

The 2012 transit of Venus

The 2012 transit of Venus was visible in many parts of the world, as shown in the diagram below, and many people watched it.

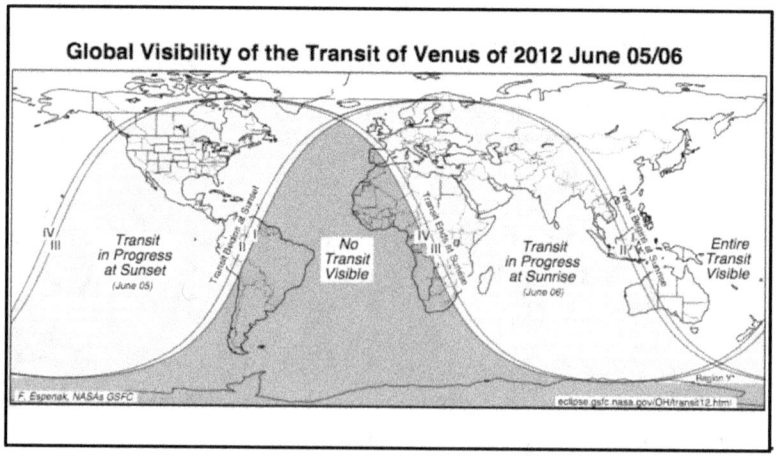

Image from NASA

The distance from the Earth to the Sun is now well known, so the 2012 transit was of lesser scientific importance than those early transits. Even so, for those people lucky enough to see it, it was an interesting sight to observe. By projecting the image of the Sun, from a telescope or a single lens from a pair of binoculars onto a piece of paper, the large slowly moving black dot on the Sun's surface could be safely seen. It was also possible to observe the transit directly through a telescope if a filter was used to cut out most of the Sun's light.

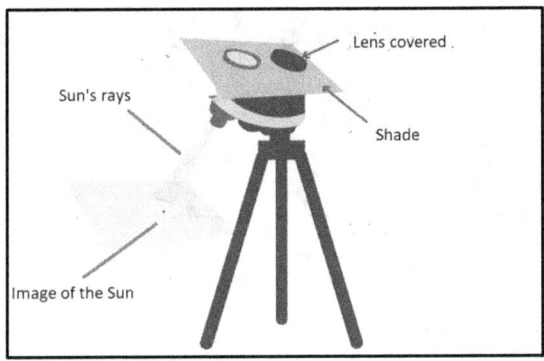

Many viewers saw the transit through the special darkened glasses which are sold to view solar eclipses.

Children Observing the 2012 transit in East Timor – image from Wikimedia commons.

Chapter 3 notes Interval between transits

The diagrams showing the orbits of the Earth and Venus have been slightly simplified. The Earth's orbit around the Sun is actually elliptical rather than circular. So, from Kepler's laws, the Earth travels in an uneven speed in its orbit, travelling faster when it is closer to the Sun. This means that, rather than there being exactly 584 days from one inferior conjunction to the next, the interval between successive inferior conjunctions varies.

Date of Inferior Conjunction (IC)	Number of days since previous IC
25/03/2017	588
26/10/2018	580
03/06/2020	586
09/01/2022	585
13/08/2023	581

The orbit of Venus around is much more circular than the Earth's. So its speed around its orbit varies less.

Chapter 4

Living on Venus

It is conceivable that one day in the distant future humans might build settlements and live on the planet Venus. Because it is well beyond what we can achieve with our current technology, it is a topic that been more in the realm of science fiction rather than factual scientific writing. However, even though there are massive obstacles in the way, I think it could conceivably happen at some point in the far distant future.

Why would we want to live on Venus?
The first thing we need to consider is why humanity would want to leave Earth and establish colonies elsewhere. Although there are no immediate plans by any nation to do this, there are a good number of reasons why this may happen eventually.

Reason 1: To ensure the continuation of humanity
While the human species is restricted to life on a single planet it is vulnerable to extinction caused by natural disasters, such as the massive impact which wiped out the dinosaurs, or man-made disasters such as an all-out nuclear war or catastrophic climate change. If humans could live in a self-supporting colony outside the Earth, then this would provide a **Plan B** to allow the continuation of our species. The British theoretical physicist Stephen Hawking (1942-2018) was a keen advocate of mankind building settlements in other parts of the Solar System and said

> *'I believe that the long-term future of the human race must be space and that it represents an important life insurance for our*

future survival, as it could prevent the disappearance of humanity by colonising other planets.'

Image from Wikimedia Commons

Reason 2: To spread human civilisation to other places

Since humans first evolved, they have constantly sought to expand to new territories and have built large cities in areas such as deserts which do not have the natural resources to support life, not even water. There seems to be an inbuilt biological imperative to find other places to live. For this reason, humans may one day extend their civilisation beyond our planet.

Reason 3: To stimulate the economy

Despite the enormous cost, building colonies outside the Earth would give a huge stimulus to the Earth's economy. There would undoubtably be huge technological challenges to overcome. However, conquering these challenges may well provide spin-offs in the same way that the American Apollo programme in the 1960s and early 1970s led to huge technological developments which were unconnected to space travel, such as advances in computing.

Reason 4 – Building international cooperation

Building large international settlements in space would almost certainly be too large an undertaking for a single nation. It is an endeavour which would need peaceful cooperation between many nations of the world. It is of course possible that, when the time arises, individual nation states may not exist and the Earth is run by single government.

All the reasons above do not just apply to Venus. They also apply to any colonies outside Earth such as large colonies in space, colonies on asteroids, colonies on moons of planets and colonies on the planets which have a solid surface (such as Mercury and Mars). Venus, however, has four advantages over the other possible places in the Solar System where humans could live.

Reason 5: Venus's proximity to Earth.

Venus is the closest planet to the Earth. At its closest approach it is 38 million km away from Earth, whereas Mars at its closest approach is 56 million km and Mercury 77 million km. This proximity makes it easier and cheaper to transfer materials from Earth to Venus than to any other planet. The only place which is closer to the Earth which humans could colonise is the Moon, but the Moon has other limitations, particularly its small size and low gravity.

Reason 6: Large Surface Area

Venus is almost the same size as the Earth. This means that it has about three times the surface area of Mars and twelve times the surface area of the Moon, giving a greater area to colonise, which would mean that it could potentially support a larger population.

Relative sizes of Venus, Earth and Mars-Images from NASA

Reason 7: Venus has a similar gravity to the Earth

When astronauts spend long periods of time in a low gravity environment, such as the International Space Station (ISS), their bones and muscles weaken. Although their muscle strength can be preserved by a strict exercise regime, nothing can be done to prevent bone loss. A strong skeleton is not needed to support a body which weighs nothing, and studies have shown that during space missions astronauts lose 1-2 % of their bone mass for each month of weightlessness. The calcium from their bones is excreted in their urine, sometimes so much calcium is lost that they develop kidney stones. The experience of the ISS astronauts would suggest that this rate of bone loss does not level off and continues at the same rate. After more than two years in low gravity, astronauts' bones are likely to be so weak they would easily fracture and would be unable to support their weight when they returned to Earth. This may be a limiting factor for how long humans can spend in a low gravity environment. The gravity on the Moon is 16% that of the Earth, and although both Mars and Mercury have stronger gravity, it is still much weaker at only 38% of the Earth's. It is not known if this would be sufficient to prevent bone loss, and it is also unclear how children born on a planet with low gravity would develop. Would their bones be so weak that they could never live on Earth later in life? The surface gravity on Venus is 91% of that of the Earth, so this would not be an issue.

Reason 8: Plenty of solar energy available

Any colony would be likely to use solar energy as its main energy source, for heating and to generate electricity. Venus is closer to the Sun than the Earth and it receives roughly twice as much solar energy. So, a given area of solar panels would be able to generate twice as much as much electricity as they would on Earth.

The following table below gives a comparison of some of the properties of Mercury, Venus, the Moon and Mars.

	Mercury	Venus	Moon	Mars
Min distance from Earth	77,300,000 km	38,200,000 km	363,000 km	55,700,000 km
Surface area	74,800,000 square km	460,000,000 square km	37,900,000 square km	145,000,000 square km
Surface Gravity Earth =1	0.38	0.91	0.16	0.38
Mean distance from Sun	0.39 AU	0.72 AU	1.0 AU	1.52 AU
Solar energy Earth=1	6.6	1.92	1	0.43

*In the table above the **Solar energy** figure is the maximum amount of energy the object receives from the Sun when the Sun is directly overhead, assuming that there is no absorption by its atmosphere.*

A very harsh environment

As discussed in chapter 2, Venus is a very inhospitable world. Its surface temperature is on average nearly 500 degrees Celsius and the atmospheric pressure is a crushing 92 times that of the Earth. No spacecraft has been able to survive for longer than about an hour on its surface without being destroyed by the intense heat and pressure.

As mentioned previously, its thick atmosphere smooths out the temperature across the planet, so there is no variation with latitude and, although the temperature drops with altitude, there is nowhere on the planet's surface which is less than 380 degrees Celsius. In addition, there is almost no water or oxygen in the atmosphere, and Venus does not have a magnetic field to protect the planet from the harmful effects of the solar wind.

Alternatives to living on the surface of Venus

The diagram below shows how the atmospheric temperature and pressure change with altitude. Interestingly at 50 km above the surface of the planet both the temperature and pressure are similar to the Earth's at sea level.

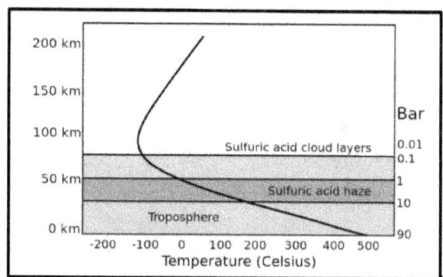

The graph above shows how the temperature and pressure of Venus's atmosphere varies with altitude - from Wikimedia Commons. 1 Bar is the atmospheric pressure at the Earth's surface.

In fact, at 50 km altitude the atmosphere of Venus is the most Earth-like environment, other than Earth itself, in the Solar System. In a paper written in 2008, the NASA scientist Geoffrey Landis suggested building floating cities in the Venusian atmosphere (Atkinson 2008). The atmosphere of Venus consists of 97% carbon dioxide, which is denser than the Earth's atmosphere. So, Landis pointed out that a large enclosed space filled with a nitrogen/oxygen mix (of similar composition to in the Earth's atmosphere) would float high above the Venusian surface in the same way that a helium balloon floats in the Earth's atmosphere.

It has been suggested by some science writers that in the future, we could explore Venus in balloons which would float above the planet at this height. It has even been proposed that we might build colonies floating in the Venusian atmosphere, should this become necessary for the survival of the human race. Inhabitants of these floating colonies could grow plants taking advantage of the strong sunlight and plentiful carbon dioxide to make oxygen and food.

Floating colonies on Venus

There are significant difficulties with this idea, but they are not insurmountable. Firstly, the clouds on Venus are made from sulphuric acid, which is highly toxic and corrosive, so any structure would have to be made from a material which was corrosion resistant. In addition, if there were a leak in the structure, leading to the lighter nitrogen/oxygen mix escaping and being replaced by heavier carbon dioxide, it would then become denser than Venus's atmosphere. The structure would then start to gradually sink. If the leak could not be repaired and it continued to sink, the structure's occupants would succumb to an unpleasant fate, being burnt and crushed as they passed through the lower atmosphere. However, if the pressure of the gases inside the structure was the same as the gases outside the rate of escape of the nitrogen/oxygen mix would be very slow. It would not be like air escaping rapidly at high pressure from a hole in a tyre or balloon. The rate of escape would be particularly slow if the size of the leak was small in relation to the size of the entire structure. This should give the occupants plenty of time to repair the leak and restore the correct gas mix, enabling it to float upwards.

It is difficult to see how any inhabitants of the floating colony would be able to leave Venus. In order to do so, a way would have to be found to get up to the escape velocity of Venus (around 37,000 km/h). On a planet's surface this can done by using a large powerful rocket, but a large heavy rocket could not be attached to a floating colony because its weight would cause the structure to sink. Possible solutions to this issue involve vast ropes and space ladders extending to well beyond Venus's atmosphere. However, the first inhabitants of the floating colony would probably have arrived on a one-way trip.

It is hard to imagine any human willingly giving up life on Earth for life enclosed in a floating space station. However, it appears that for some people the prospect of unimaginable adventure does outweigh all the disadvantages of leaving our planet. In 2013 a small Dutch company called Mars One (www.mars-one.com) advertised for twenty people who would be willing to be sent to Mars with no chance of ever returning to Earth, an astonishing four thousand people put themselves forward and paid an application fee of around $100.

In reality, it seems more likely that humans would only live on Venus in significant numbers if the entire planet could be transformed to make it more like the Earth, a process known as terraforming, the subject of the next chapter.

Chapter 5

Terraforming Venus

Terraforming is the process of changing the global environment of a planet in such a way as to make it suitable for human habitation. Because it is so far beyond our current technological capabilities, most articles about terraforming have been written by science fiction writers rather than scientists. For example, there is an entry for a terraformed Venus in the science fiction work 'Encyclopaedia Galactica', (*http://www.orionsarm.com/eg-article/48e7ce77477af*) which is set thousands of years in the future.

> *Originally a hot dry greenhouse world ... with an atmosphere consisting mostly of carbon dioxide with a surface pressure 94 times greater than that of Earth. The planet was shrouded with clouds of sulphuric and hydrochloric acid and the mean surface temperature was 480 C, making the world extremely hostile to terragen [human] and carbon-based life. Because of this the planet was sparsely populated for many thousands of years; recently it has been successfully terraformed.*

In the **Current Era** section, it states.

> *With terraforming recently completed, the Venusian surface is covered in a beautiful saltwater ocean punctuated by a number of continents and islands with a wide variety of climates. The flora and fauna have been tailored to closely resemble that of Old Earth, with some adapted xenobiotic species included. While there are*

several important population centers, most of the surface population live in small diffused settlements. Water craft are ubiquitous and range in size from a few meters to housing entire cities. Currently, only 2 billion physical inhabitants permanently live on the surface and the Solar Organization is taking steps to prevent overcrowding.

So far there has been very little detailed scientific research on terraforming Venus. Most papers on terraforming have focussed on Mars. Even so, there is no reason why - given enough resources and time - an advanced human civilisation would not be able to terraform Venus.

It has to be said at this point that the international willingness to commit to a project like this is much harder to imagine than the existence of the technical ability to do so. There would have to be fundamental changes in the way in which human beings work together on a global scale, and the conditions of life for the whole of humanity would have to be considerably better than they are now, were such vast amounts of time and money to be committed to a project like this.

However, setting that aside, it is interesting to consider the various challenges to making Venus habitable, so that humans can live and work on the planet without any need for protective equipment, such as space suits or oxygen supplies.

*What a terraformed Venus might look like
- image from Wikimedia Commons*

Reducing the high temperatures and pressures

The surface temperature of Venus is around 500 degrees Celsius and the atmospheric pressure is a crushing 92 times that of the Earth. The atmosphere consists of 97% carbon dioxide (Williams 2015) which, as discussed in chapter 2, is a powerful greenhouse gas which traps the Sun's heat. To make the planet's surface planet habitable the temperature would need to be lowered to around 0 to 35 degrees Celsius and the atmospheric pressure to a similar value to that on Earth.

One way to cool Venus and reduce its enormous atmospheric pressure would be to build a giant sunshade to block most of the Sun's rays from hitting the planet. This idea was described in detail in a paper written in 1991 by the British science writer Paul Birch (1956-2012).

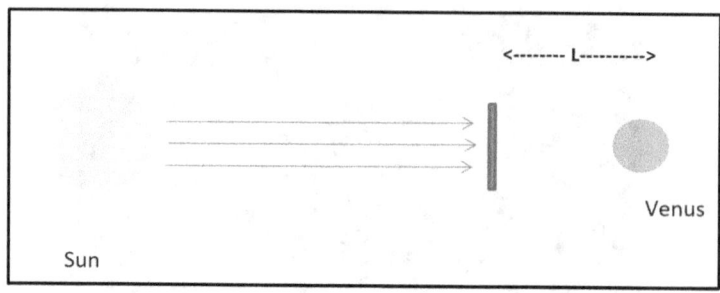

The shade would orbit the Sun at a specific point about 1 million km above Venus's surface, called the L1 Lagrange point, shown as **L** in the diagram above. At this position the shade would orbit the Sun in the same time that it takes Venus to orbit the Sun, 224.7 days, thus providing permanent shade.

The shade would need to be slightly larger than the diameter of Venus, 12,100 km, to completely block the Sun. The cost and technological challenge of building such a shade would be enormous. It would need to be 100 billion times larger in surface area than the International Space Station, shown below, which is the largest object ever built in space.

The International Space Station - Image from NASA

It would be almost impossible to get such a massive object into space in a single go. So, the shade would be built up from thousands, or more likely millions, of smaller individual shades and would take many decades to complete from start to finish. As the shade neared completion, and most of the Sun's rays were blocked from hitting the planet, the surface of Venus would begin to cool. Interestingly, Venus would then no longer be lit up by the Sun. So, to a viewer on Earth, it would go from being the third brightest natural object in the sky (after the Sun and the Moon) to being invisible. Although the shade itself would be visible from Earth.

Birch estimated that after about 100 years without sunlight the temperature of Venus would drop to 31 degrees, but clearly this timescale is only approximate. At this temperature, known as the critical point of carbon dioxide, some of the carbon dioxide in the atmosphere would start to condense from gas to liquid and the low-lying areas of the surface of Venus would begin to be covered in seas and oceans of liquid carbon dioxide. As it condensed into liquid, the amount of carbon dioxide left in the atmosphere would fall and with it the atmospheric pressure. Eventually the temperature would drop to the freezing point of carbon dioxide (-57 degrees) and the seas, oceans and lakes of liquid carbon dioxide would begin to freeze. Much of the remaining carbon dioxide in the atmosphere would fall as snow.

This entire cooling process would take hundreds of years from start to finish. The next step would be to ensure that when the shade was removed, allowing the planet to graduallywarm up, the frozen carbon dioxide would not be released back into the atmosphere, causing the temperature to rise again due to the greenhouse effect. One way this could be achieved would be to cover up the frozen carbon dioxide oceans with a thick layer of insulating material and provide some sort of refrigeration system to keep them cool. Once the carbon dioxide was safely locked away the sunshade could then be removed to allow the planet to warm up again.
Because nearly all the carbon dioxide would have been removed from the atmosphere it would no longer provide such a powerful greenhouse gas. By removing part of the shade, the temperature could be tuned to the optimum value.

Getting water to Venus

Water is essential for all forms of life found on Earth. However, Venus is an extremely dry planet: its atmosphere contains only a small trace of water and there is no water on its surface. By comparison, 71% of the Earth's surface is covered by water and there are about 1.39 billion cubic kilometres of water on the planet. The breakdown of how this water is distributed is shown below.

Water source	Water volume, in cubic kilometers	% of total
Oceans, Seas, & Bays	1,338,000,000	96.54
Ice caps, Glaciers, & Permanent Snow	24,064,000	1.74
Groundwater	23,400,000	1.69
Fresh	10,530,000	0.76
Saline	12,870,000	0.93
Soil Moisture	16,500	0.001
Ground Ice & Permafrost	300,000	0.022
Lakes	176,400	0.013
Fresh	91,000	0.007
Saline	85,400	0.006
Atmosphere	12,900	0.001
Swamp Water	11,470	0.0008
Rivers	2,120	0.0002
Biological Water	1,120	0.0001

Source: U.S. Department of the Interior 2015

It would undoubtedly be necessary to import a huge amount of water to Venus to make the entire planet habitable for plant and animal life, but it would not be necessary to have most of the planet covered with deep oceans. Around 30 million cubic kilometres of water (roughly 2% of the amount found on Earth) ought to be sufficient.

There various ways of transporting this water to Venus. One option would be to transport the water from the seas and oceans of Earth by cargo-carrying spacecraft. However, one cubic kilometre of water weighs one billion tonnes, so 30 million cubic kilometres of water would weigh a massive 30 thousand trillion (30,000,000,000,000,000) tonnes. If we had a massive fleet of one thousand spacecraft, each of which could carry a payload of ten thousand tonnes would need 3 billion missions to carry this amount of material. This is clearly not feasible.

Another possibility would be to transport only hydrogen, and not necessarily from the Earth. A water molecule consists of two hydrogen atoms and one atom of oxygen, but the latter is sixteen times heavier than a hydrogen atom. The mass of material taken to Venus would therefore be reduced by 90%. Water could be produced by chemical reaction of the transported hydrogen with the remaining carbon dioxide in Venus's atmosphere. The hydrogen could be created on Earth by splitting water molecules into their component elements, or it could be sourced elsewhere. For example, it could be collected by scooping it up on an orbiting ring from the atmosphere of one of the giant planets in the outer Solar System and transporting it to Venus by spacecraft. Regardless of where the hydrogen is sourced, this method of providing Venus with water would still be prohibitively expensive.

A fascinating alternative way of getting all the water needed to Venus was suggested in Birch's paper. It involves moving one of Saturn's ice moons into orbit around Venus and then breaking it up, releasing its water onto the planet.

Saturn has a number of ice moons such as Hyperion (shown below), an irregularly shaped object 360 km by 260 km which consists mainly of ice covered in a thin layer of rock.

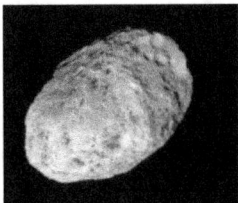

Image from NASA

Birch's proposal was that we build a huge structure on Hyperion which would use the Sun's heat, concentrated by mirrors, to put out a jet of steam into space in the same direction as Hyperion orbits Saturn. This jet of steam would provide a force which would gradually cause the moon to lose energy and move into a lower orbit, as shown in the diagram below:

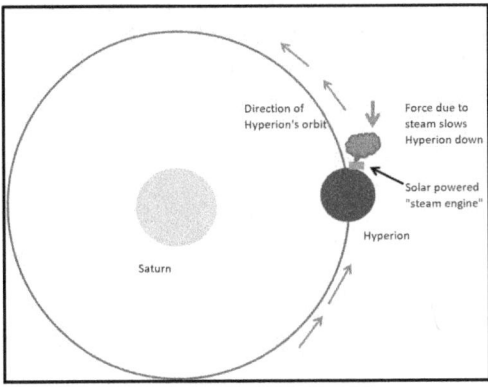

The shape and distance from Saturn of Hyperion's new orbit could be finely tuned by adjusting the force provided by the solar powered steam engine. Birch's proposal was to place Hyperion into an elliptical orbit in which it passed close to Saturn's giant moon Titan. He calculated that to move Hyperion into this new orbit would take about 30 years.

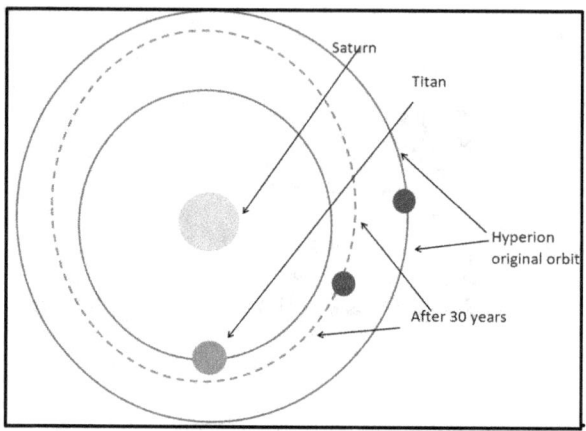

Eventually, Hyperion will pass close to Titan in its new orbit and will be affected by its gravity which will pull it towards Titan. This action will cause Hyperion to accelerate to such a velocity – around 28,000 km/h - that it can escape from Saturn. This technique is known as a gravitational slingshot and NASA mission planners commonly use it to reduce the amount of fuel needed when they send spacecraft to the outer Solar System.

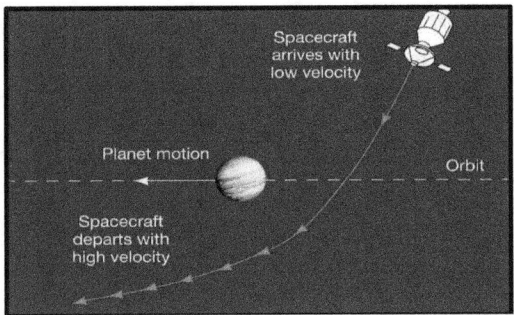

A gravitational slingshot - as the spacecraft approaches the planet, the planet's gravity speeds it up and changes its trajectory

If the speed and angle of approach to Titan were just right, after Hyperion escaped from its orbit around Saturn it would be on a path which took it close to the giant planet Jupiter.

As Hyperion passed Jupiter, the giant planet's gravity would perform a second gravitational slingshot to change its trajectory and hurl it into the inner Solar System. If the angle of approach to Jupiter were correct, Hyperion would be placed on a path on which it passed Venus slowly enough to be captured by the planet's gravity. Hyperion would then go into orbit around Venus. When this had been achieved the heat of the Sun would cause the ice to start to melt and a way would have to be found to transfer the water from Hyperion to Venus's surface.

Although this somewhat convoluted plan might appear to be something out of an exotic science fiction story, it obeys all the laws of physics and could potentially be achieved by an advanced human civilisation which was able and willing to devote enough resources to do it. However, building the solar-powered steam engine structure on Hyperion would be a massive feat of engineering which would almost certainly take decades and cost trillions of dollars in today's money.

Adding oxygen to Venus's atmosphere

In order to be habitable by animal life, including humans, Venus would need a similar level of oxygen in its atmosphere to that on the Earth. Oxygen makes up about 21% of the atmosphere on Earth, whereas Venus's atmosphere has almost no oxygen. This could be relatively easy to resolve, compared to the other challenges, if a large amount of plant life were able to thrive in the Venusian environment. Through photosynthesis, the plants would use the energy from the Sun to convert carbon dioxide and water into carbohydrates (such as cellulose and starch) and oxygen. This would increase the oxygen concentration in the atmosphere.

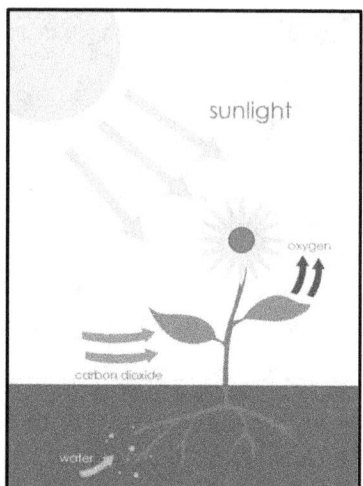

Image from Wikimedia Commons

Overcoming the long day/night cycle

On Venus the slow rotation of the planet means that there are the equivalent of 58.4 Earth days of continuous daylight followed by 58.4 days of darkness. On Earth life has evolved around a 24-hour day and most lifeforms would struggle to adapt to such a long day/night cycle. A shorter cycle could be achieved by having a system of sunshades and mirrors in orbit around the planet, although this would obviously be a huge and expensive project.

Overcoming the lack of a magnetic field

The Earth's magnetic field forms a shield around the planet which protects its surface from energetic electrically charged particles from the Sun (known as the solar wind) and from outer space (cosmic rays). Without a magnetic field there would be an increased risk of cancer for anyone who ventured outdoors, although it is difficult to predict exactly much exposure to the solar wind and cosmic rays would present a threat to health. To make Venus completely habitable, so that its inhabitants could live and work safely outdoors, it would need to be given an artificial magnetic field. The next two chapters will discuss how this could be achieved.

Chapter 6

Planetary Magnetic Fields

The Earth is unique among the four inner planets in our Solar System (Mercury, Venus, Earth and Mars) in having a strong magnetic field. As mentioned in the previous chapter, it protects us from harmful radiation from space, but this invisible field, which causes the needle of a compass to point North, also enabled navigators to find their way across the sea for centuries and is used by some birds and land animals in their migratory patterns.

Overview of the Earth's magnetic field

As an approximation, the Earth's magnetic field behaves as though there were a giant bar magnet inside the planet. The poles of this invisible magnet, marked as **Nmag** and **Smag** in the diagram below, lie close to the geographical poles marked as **N** and **S**.

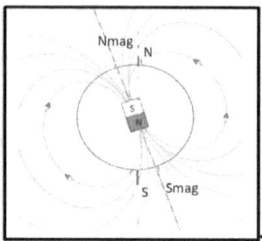

In reality, treating the Earth's magnetic field as a simple bar magnet is an over-simplification. The true nature of the Earth's magnetic field is more complex than this; there are many places known as magnetic anomalies where the Earth's magnetic field is stronger or weaker than it would be with this simple model.

The strength of a magnetic field is usually measured in teslas, named after the Serbian-American physicist, engineer and inventor Nikola Tesla (1856-1943), pictured below. A magnetic field of 1 tesla is relatively strong. So, for instance, a small bar magnet has a field strength of around 0.01 tesla. The Earth's magnetic field is much weaker than this. It varies between 30-65 millionths of a tesla, usually denoted as 30-65 microtesla, and is stronger near the magnetic poles and weaker near the equator.

Nikola Tesla - Image from Wikimedia commons

What causes the Earth's magnetic field?

The generally accepted theory is known as the dynamo theory. In summary, it states that the Earth's magnetic field is generated by movements stirred up in its outer core by the Earth's rotation, known as convection currents. The Earth's outer core is made of iron which is a good conductor of electricity and it is molten because of the high temperatures. Following this theory, for any planet to have a magnetic field, part of its interior must consist of a liquid which conducts electricity and it must be rotating rapidly enough to generate convection currents.

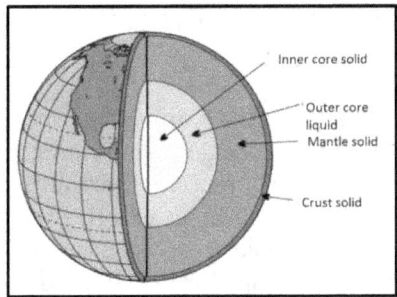

Magnetic fields of the other inner planets

- **Mars** has no magnetic field. The reason for this is because it is much smaller than the Earth and its smaller size means that in the time since the planets were formed, roughly 4.5 billion years ago, its inside has cooled down more than the Earth's. In fact, Mars's interior has cooled down so much so that it is solid and so there can be no convection currents to create a magnetic field.
- **Venus** is the same size as the Earth and is of similar internal composition with a liquid outer core. It does not have a permanent global planet-wide magnetic field because it rotates far too slowly to create convection currents in its interior. However, the interaction between the electrically charged particles in the solar wind and Venus's atmosphere induces an extremely weak and variable magnetic field which can be as strong as 0.15 microtesla in places (Lumann and Russell 1997).

- **Mercury**, the innermost planet, has a weak permanent magnetic field, around 1% of the strength of the Earth's field. The fact that any magnetic field exists is surprising, and still not fully understood. Prior to 1974, astronomers generally assumed that Mercury had no magnetic field for the same reason as Mars - that its small size meant that the inside of the planet had cooled down so much that its entire core was solid. Plus, like Venus, Mercury rotates very slowly, once every 59 Earth days, which means that even if the outer core were still liquid its rotation would be too slow to create any magnetic field. However, the visit of the Mariner 10 space probe to the planet in 1974 showed that the assumption was wrong, and astronomers are still unclear as to why the magnetic field exists. Perhaps the core is still liquid, or it may be that there is some sort of solidified permanent magnet within a solid core

Effect of the lack of magnetic field on Venus

As discussed in the previous chapter, the terraforming of Venus would give it a similar temperature and atmospheric pressure to that which currently exists on the Earth. However, a lack of a strong global magnetic field like that found on Earth would cause significant obstacles to humans settling on Venus. The Earth's magnetic field forms a protective shield called the magnetosphere protecting us from a stream of electrically charged particles from the Sun called the solar wind. This is shown in the diagram below.

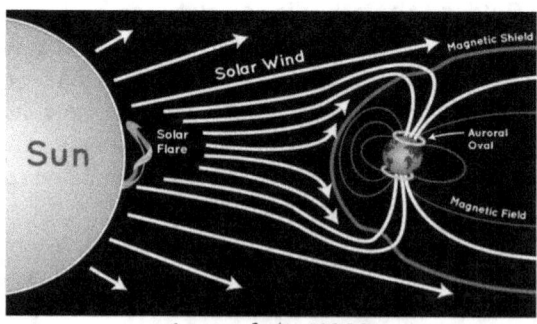

Image from NASA

Venus has virtually no magnetosphere because it has no significant magnetic field. One effect of this is that its atmosphere is being slowly eroded as the outer layers of gas are blown away into space by the solar wind. This erosion is a very slow process and it would take many millions of years for a terraformed Venus, with a similar atmospheric density to the Earth, to lose a significant portion of its atmosphere. It would therefore not be an obstacle to settlement because the fraction of atmosphere lost would be negligible over the lifetime of a human being.

A more serious issue is that without the protective magnetosphere the solar wind would prevent an ozone layer being created. On Earth there is a region of the atmosphere around 35 km in altitude where the concentration of ozone gas is at its greatest (NASA 2013).

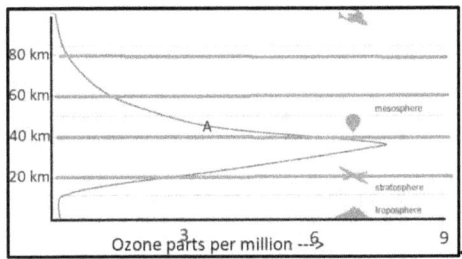

The light blue line labelled A shows how the ozone concentration in the atmosphere varies with altitude- Data from NASA

This ozone layer prevents most of the Sun's harmful ultraviolet rays from hitting the Earth's surface. On a terraformed Venus without a magnetosphere the electrically charged particles in the solar wind would break up any ozone formed in the upper atmosphere. Without a protective ozone layer, the high levels of ultraviolet (UV) radiation hitting the planet's surface would mean that anyone who ventured outdoors on Venus would be exposed to a high risk of skin cancer. Lack of a protective ozone layer would also prevent agriculture outdoors, as prolonged exposure to UV radiation would break up many of the complex organic molecules found in living organisms.

Another risk to health is that many of the particles from the solar wind and other deadly particles from space, known as cosmic rays, would hit the planet's surface, exposing its inhabitants to a serious risk of ill health, as cosmic rays are known to affect the process of cell division. People would therefore be at risk of cancer and growth disorders, but information about this is sketchy, because the only people to have ventured outside the Earth's magnetosphere were the Apollo astronauts in the late 1960s and the early 1970s. Most of them reported seeing flashes of light even when their eyes were closed: this was due to cosmic rays passing through their spacesuit and their bodies and being seen as a flash of light as they hit the back of their eyes. Some of them later developed cataracts at an earlier age than would normally be expected. The Apollo astronauts could have received a lethal dose of radiation during their voyages if there had there been a solar storm, when the radiation from the Sun is much more intense. To make Venus fully habitable it would therefore be necessary to give it an artificial magnetic field, and how this could be done is discussed in the next chapter.

Chapter 7

Giving Venus a Magnetic Field

As discussed in the previous chapter, on a terraformed Venus, humans would not be able to venture outdoors without protective clothing unless the planet had a strong planet wide magnetic field.

Creating a magnetic field
In order to see how Venus could be given such a field, it is first necessary to understand some of the principles of electromagnetism. When an electric current flows though a loop of wire it induces a magnetic field.

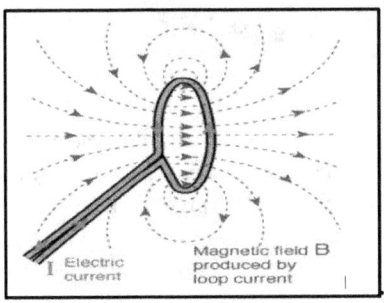

The strength of this magnetic field is directly proportional to the strength of the electric current through the wire and inversely proportional to the diameter of the loop. This inverse dependence on the diameter of the loop means that if we were to pass a current of 1000 amps (which is the typical current flowing through high voltage power lines) through a massive metal ring which was the same diameter as Venus (12,100 km), because of the immense size of this loop the magnetic field generated would be extremely weak – about one million times weaker than the Earth's field. To generate a planet-wide magnetic field of a similar strength to the Earth's field, there would need to be a gigantic power-generating complex on the planet which could pass a current of 1 billion amps through the planet-size loop.

This would be an enormous current to keep flowing through such an immense structure. We could reduce the amount current needed by creating a coil or a series of separate rings around the planet and the magnetic field from each individual ring would add up to give the total magnetic field. If a future civilisation were to construct a planet-sized structure of 10,000 separate rings, then they would still need to pass a current of **100,000 amps** through each ring. The main problem they would face in trying to generate a magnetic field this way is that it would consume an enormous amount of electric power. The power used when an electric current flows is given by the equation:

$$P = I^2 R$$

Where

>P is the power - measured in watts
>I is the electric current - measured in amps
>R is the resistance: this is a measure of the extent to which a material resists the flow of electricity and is measured in units called ohms.

So, using this equation, to reduce the power needed, the resistance needs to be as low as possible. Wires and cables made from a good conductor of electricity such as copper have a low resistance and the thicker the wire the lower its resistance. If each of the 10,000 planet sized rings were made from a copper cable 20 cm thick, then the resistance of a metre length of the cable would be very low, only 0.000 000 54 Ohms *(see note 1)*. Even so, if a current of 100,000 amps were passed through this thick cable, it would use 5.4 kilowatts of power which would be dissipated as heat.

However, this is only the heat generated per single metre of cable. The total length of the 10,000 rings wrapped around Venus would be 400 billion metres. So, the whole structure would use an incredible 2,000 trillion watts of power. Maintaining an artificial magnetic field in this way for a year would need 18 million trillion watt-hours of energy. This is roughly 800 times larger than the entire Earth's electricity consumption in 2017 (Enerdata 2018). If this amount of energy were to be generated by solar power, an area of roughly 7 million square km, which is roughly 70% of the area of the US, would need to be covered with solar cells *(see note 2)*.

Not only would there be challenges in generating this amount of electrical energy, it would also be a major problem to dissipate the heat generated, to prevent the copper coils getting so hot that they would melt.

Superconductors

Another option would be to build the coils out of a superconducting material. Superconductors were discovered in the early twentieth century and have many applications including maintaining the high electric currents needed to produce strong magnetic fields in medical devices such as MRI scanners and particle accelerators such as the Large Hadron Collider.

Superconductors are materials which have zero electrical resistance and no power is used when a current flows, because **P= I^2R** (and R=0). An electric current can flow through a superconducting loop indefinitely without diminishing. However, all known superconductors only superconduct (that is, have zero resistance) below a certain temperature which is normally very low. They also have a critical current above which the material no longer superconducts. This critical current depends upon the material the superconductor is made from, its thickness and the temperature. For example, the critical current of a 1.8 cm diameter tube made from the metal tin at a temperature of -271 degrees C is 2000 amps. If a planet-sized ring were made from a superconductor, although there would be no power lost due to electrical resistance, a significant amount of power would need to be supplied to keep the structure very cold. If the cooling failed, the material would no longer be superconducting and the magnetic field would disappear as the electric current faded away. Building such a massive structure of refrigerated planet sized rings would obviously present immense challenges, and although a very advanced civilisation may well be able to overcome these challenges, alternative and simpler solutions might one day be possible.

For example, new higher temperature superconductors might be discovered. When the first superconductors were identified, they only became superconducting close to absolute zero (-273.15 degrees Celsius), which is the lowest possible temperature. In the course of time new materials were discovered which superconducted at higher temperatures. However, most of these higher temperature superconductors do not support high currents and still need to be cooled to around -200 degrees Celsius. A real breakthrough was made in 2015 when it was discovered that hydrogen sulphide, the gas responsible for the bad smell in rotten eggs, became superconducting when cooled to -70 degrees, although it must be compressed to a pressure of 1 million atmospheres (Drozdov et al 2015). This amazing discovery raises the possibility that in the future we will have room temperature superconductors, albeit at very high pressures, and perhaps these materials could be used to build a superconducting ring around Venus.

Speeding up the planet's rotation

As discussed in the previous chapter, the reason that Venus does not have a strong planet-wide magnetic field is because it rotates so slowly that it does not generate convection currents in its interior. As an alternative to generating a magnetic field from a strong electric current, another option would be to speed up the planet's rotation. If Venus were rotating at the same rate as the Earth, then our current theories on the origin of the Earth's magnetic field suggest that this rotation would induce a magnetic field of a similar strength to the Earth's.

In reality, however, changing the rotation speed of a planet the size of Venus would be such a massive feat of macro-engineering on a planetary scale that generating the magnetic field using superconductors would probably be more feasible.

Chapter 7 Notes

Resistance and resistivity

To work out the resistance of a given length of copper cable we need to know the **resistivity** of copper. Resistivity is a fundamental property of a material and does not depend on its size or shape (although it does vary with temperature). It is measured in units called ohm-metres, and is usually given the symbol ρ (the Greek letter rho). It is the resistance of a block of a material 1 square metre in diameter and one metre in length.

As shown in the diagram below, the resistance **R** in ohms of a piece of material of resistivity ρ, cross sectional area **A** (measured in square metres) and length **L** (measured in metres) is given by

$$R = \rho L / A$$

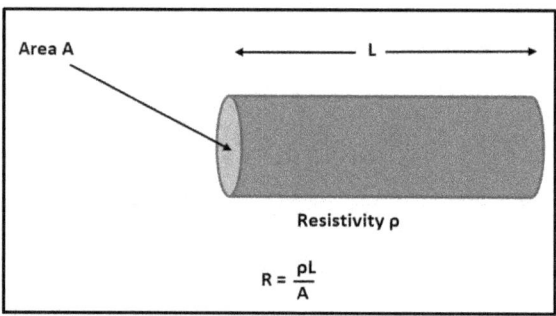

If the resistivity of copper is 17 nano-ohm metres, a thick cable 20 cm in diameter has a cross sectional area of 0.0314 square metres. so the resistance of a one metre length of the cable is 0.54 micro-ohms.

Solar energy on Venus

The calculation assumes that the average amount of solar radiation hitting a point on Venus's surface over the course of a Venusian year is 700 watts per square metre, allowing for variations with latitude and the day-night cycle. It also assumes that solar panels can be mass-produced with an energy efficiency of 40%, which is much higher efficiency than today's solar panels, typically 10-15%.

Chapter 8

Conclusions

It is remarkable how much our knowledge about Venus has advanced in the last sixty years. The planet appears as a featureless disc through even the largest telescopes, so astronomers knew very little about it until as late as 1960. Properties such as its surface temperature and pressure were unknown, and its rotation period could not be determined as there were no surface features to be observed. In the 1950s some thought that conditions might be similar to those on Earth, and that even if alien life did not live there, it was a planet which humans would explore in the near future and build bases on.

Today we know that Venus is a harsh hostile world, having extremely high temperatures and crushing pressures. It could not possibly support any form of life. Unlike Mars, the Moon or Mercury, it would be impossible for humans to land anywhere on its surface and survive, even wearing a space suit. After landing on the surface, no spacecraft has ever transmitted any data for longer than one hour before being destroyed by the high temperatures and pressures.

Nevertheless, for the reasons I discussed in Chapter five, I think that eventually, in the far distant future, humanity could conceivably terraform Venus to make it suitable for habitation. Perhaps 10,000 years from now, a terraformed Venus could support a population of billions of people, given its large size and abundant solar energy. Maybe one day an inhabitant of Venus will gaze into the night sky and see a bright blue-green planet, the brightest object in the sky, bright enough to cast faint shadows, large enough to be seen as a disc than a point of light. This will be the Earth, the planet their ancestors came from, perhaps where other of their ancestors still live.

Steve Hurley

October 2019

Appendix I: Glossary

Term	Definition
Akatsuki	Japanese space probe launched in 2010 to study the atmosphere and climate of Venus. It started the main scientific part of its mission in 2016.
albedo	A measure of the amount of radiation which is reflected by a body. It is measured on a scale between 0 and 1. • An albedo of 0 means that the object absorbs all radiation hitting it and reflects none back, so such an object would appear completely black. • An albedo of 1 means that the object reflects all radiation back. A planet entirely covered in fresh snow would have an albedo close to 1.
aphelion	The point in the orbit of a planet when it is furthest away from the Sun.
astronomical unit	Unit of distance used in astronomy. It is normally abbreviated to AU and is equal to 149,597,870 kilometres. One AU is equal to the mean of the Earth's aphelion and perihelion.
elongation	How far away a planet appears from the Sun in the sky. This is measured in degrees. • An elongation of zero degrees means that the planet is directly in front of or behind the Sun.

Term	Definition
	• An elongation of 180 degrees means that the planet is in the opposite direction from the Sun. The maximum elongation of Venus is 47.8 degrees. The light green line shows the direction of the Sun. Each of the dark green lines shows the direction of Venus at various points in its orbit. The angle between a dark green line and the light green line is the elongation at the given point.
geocentrism	Obsolete astronomical theory in which the Earth is the centre of the Universe and the stars, planets, Sun and Moon are in orbit around it.
heliocentrism	Theory associated with Nicolas Copernicus in which the planets orbit the Sun.
inferior conjunction	Position in the orbit of a planet when it is directly between the Earth and the Sun. Only planets which orbit inside the Earth's orbit, that is Mercury and Venus, can have inferior conjunctions.

inner planet	This term can have two different meanings.
	   1) One of the four small rocky planets Mercury, Venus, Earth and Mars, which are in the inner region of the Solar System.
	   2) A planet which is closer to the Sun than the Earth, in this case only Mercury and Venus.

Kepler's Laws	Three laws discovered by the German mathematician and astronomer Johannes Keppler (1571-1630), which govern the way that a satellite moves around a larger body.

Term	Definition
	Although Kepler formulated his laws in terms of the planets orbiting the Sun, they apply to any smaller body orbiting a larger more massive body. **Kepler's first law** states that the planets move in ellipses with the Sun at one of the focuses of the ellipse. **Kepler's second law** states that a line drawn between the Sun and a planet sweeps out equal areas at equal times. **Kepler's third law** states that the square of the orbital period of a planet is directly proportional to the cube of the semi-major axis of its orbit. Kepler's laws are discussed in more detail in chapter 3.
Magellan	American spacecraft launched in 1989 which mapped Venus by radar between 1989 and 1994.
magnetosphere	The protective shield around a planet produced by its magnetic field, which prevents the solar wind and cosmic rays hitting its surface and lower atmosphere.
magnitude	A measure of how bright an astronomical object appears when viewed from Earth. The lower the value of the magnitude, the brighter the object. A star with a magnitude of +6 is just bright enough for someone with good eyesight to see without a telescope, against a dark sky. The brightest star in the sky, Sirius, has a magnitude of -1.5. At the brightest point in its orbit Venus has a magnitude of around -4.5. (Because of its orbit is elliptical it sometimes can get brighter than this). This makes Venus the third brightest object in the sky after the Sun (magnitude -26.7) and the Moon (magnitude -12.7). Magnitudes are normally measured in visible light at a wavelength of 540 nanometres.
Mariner	A series of unmanned American space probes launched between 1962 and 1973. In 1962, Mariner 2 became the first spacecraft to explore Venus.
nodes	The orbits of all the planets are tilted at different angles with respect to the Earth's orbit. The two places in the orbit of a planet where it intersects the plane of the Earth's orbit are known as the nodes. These are marked

Term	Definition
	as A and B in the diagram below.
opposition	Position in the orbit of a planet when it is in the opposite direction from the Sun as seen from the Earth. Only planets outside the Earth's orbit can be in opposition.
perihelion	The point in the orbit of a planet when it is closest to the Sun.
semi-major axis	Half the length of the longest axis of an ellipse. For a planet orbiting the Sun the semi-major axis is equal to the mean of the perihelion and the aphelion.
solar constant	The mean amount of radiation from the Sun that is received at the Earth's mean distance from it (1 AU). The solar constant is a not a true constant because the Sun fluctuates slightly in brightness. It has a value of 1.361 kilowatts per square meter (kW/m²) at solar minima and is approximately 0.1% greater at solar maxima.
synodic period	The period of time it takes a planet moving inside the Earth's orbit to complete one extra orbit and line up with the Earth again **OR** The period of time it takes a planet moving outside the Earth's orbit to complete one fewer orbit and line up with the Earth again. The synodic period of Venus is 584 days.
terraforming	The process of changing the global environment of a planet in such a way as to make it suitable for human habitation. On a terraformed planet, humans can live and work on its surface without any specialised life support equipment.
transit	When a smaller astronomical body (for example Venus) passes in front of a larger astronomical body (for example the Sun).
Venera	A series of unmanned Soviet space probes launched between 1961 and 1983. In 1970 Venera 7 became the first spacecraft to successfully land on another planet.
Venus Express	European space probe which orbited Venus between 2006 and 2015. Its mission was to study the atmosphere

Term	Definition
	of Venus.

Appendix II: Determining the size of the AU

Imagine two observers watching a transit of Venus. These two observers are located at points A and B on the Earth's surface.

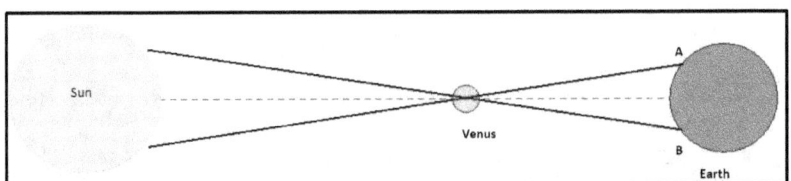

The two observers will see Venus follow different paths across the Sun's surface during a transit. In general, the transit may also start and finish at different times for the two observers.

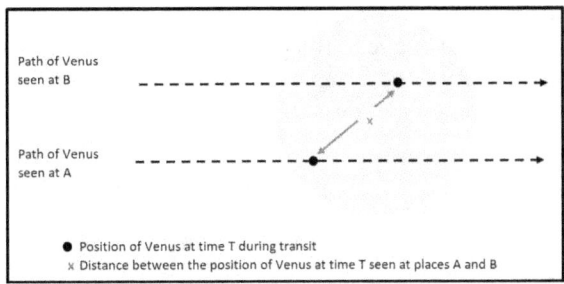

Trigonometry can be used to calculate the distance between Earth and Venus D_{EV}, as shown in the diagram below.

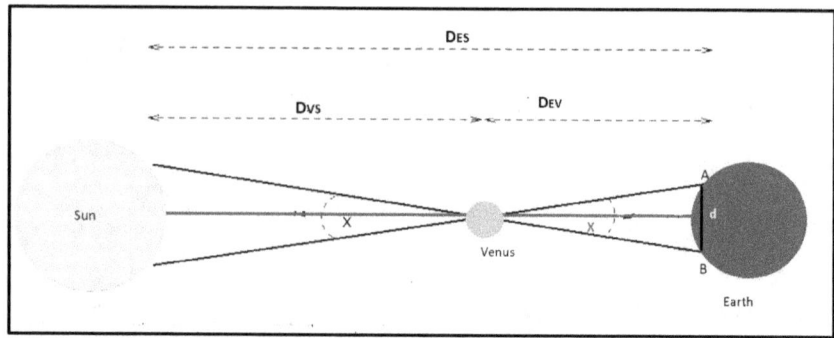

In this diagram the following abbreviations have been used:
- D_{ES} is the distance between the Earth and the Sun. This has a value of one astronomical unit (AU), because the astronomical unit is defined as the mean distance between the Earth and the Sun. In the early seventeenth century its actual value in miles was unknown.
- D_{VS} is the distance between Venus and the Sun. This has a value of 0.72 AU.
- D_{EV} is the distance between the Earth and Venus. This has a value of 0.28 AU.
- The angle **X** is the difference between the position of Venus on the Sun's disc when viewed from points A and B. This can be measured by observing the transit.
- d is the distance between points A and B, which is known.

Then from trigonometry
$$\text{Tan}(0.5x) = 0.5d / D_{EV}$$
Therefore, the actual distance between the Earth and Venus D_{EV}, which is equal to 0.28 AU, could be calculated in real units such as miles. Once this value was known, the size of the astronomical unit could be determined by dividing by 0.28.

Appendix III: Venus Facts

The tables in this section provide extra data on Venus. The data has been taken from the NASA factsheets (Espinak 2017).

Table I Bulk parameters

	Venus	Earth	Ratio (V/E)	Notes
Mass (10^{24} kg)	4.8675	5.9724	0.815	
Volume (10^{10} km^3)	92.843	108.321	0.857	
Equatorial radius (km)	6051.8	6378.1	0.949	
Polar radius (km)	6051.8	6356.8	0.952	
Mean radius (km)	6051.8	6371.0	0.950	
Ellipticity (flattening)	0.000	0.00335	0.0	1
Mean density (kg/m^3)	5243	5514	0.951	
Surface gravity (m/s^2)	8.87	9.80	0.905	
Escape velocity (km/s)	10.36	11.19	0.926	
Bond albedo	0.77	0.306	2.52	2
Geometric albedo	0.689	0.434	1.59	2
Solar constant (W/m^2)	2601.3	1361.0	1.911	3
Topographic range (km)	13	20	0.650	4
Number of natural satellites	0	1		

Notes:

1. **Ellipticity** is a measure of how flattened a body is and is defined as its equatorial diameter divided by its polar diameter.

2. The **albedo** is a measure of the fraction of the TOTAL radiation hitting a body which is reflected back into space.

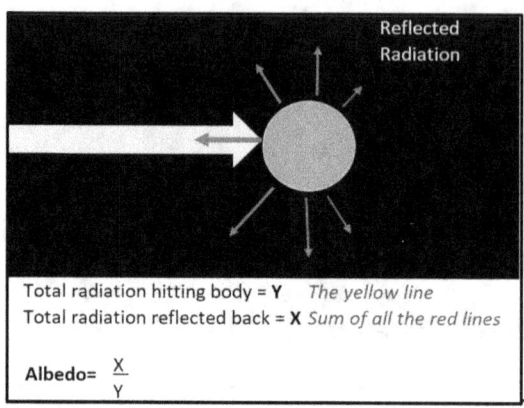

Total radiation hitting body = Y   *The yellow line*
Total radiation reflected back = X   *Sum of all the red lines*

Albedo = $\frac{X}{Y}$

It is measured on a scale between 0 and 1.
- An albedo of 0 means that the object absorbs all the radiation hitting the body and reflects none back, such an object would appear black.
- An albedo of 1 means that the object reflects all radiation back. A planet whose surface was covered in snow would have an albedo close to 1.

Strictly speaking the definition above is for what is known as the **Bond albedo**. However, the Bond albedo cannot be easily measured because, from Earth, we cannot detect the amount of radiation reflected in all directions, only in the direction facing Earth. There is another type of albedo commonly used known as the **geometric albedo** which is the amount of radiation reflected back, compared to an idealised flat reflecting disk. when the planet in a direct line with the Sun, e.g. when a planet is at opposition or the Moon at full Moon.

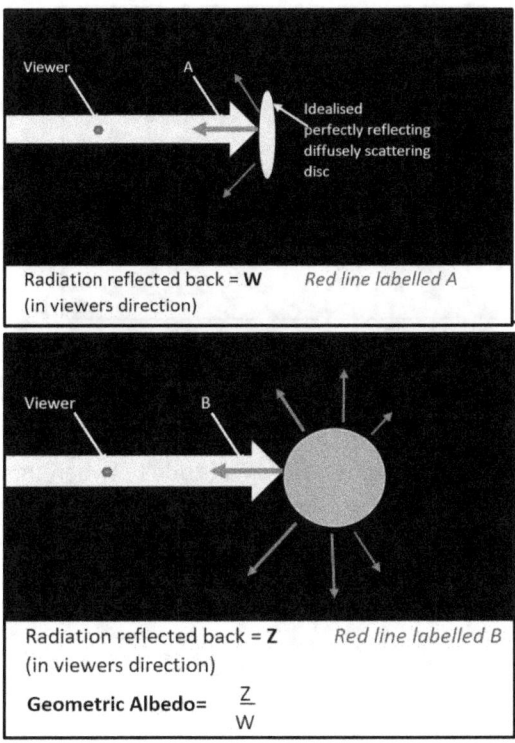

3. The solar constant is the strength of the radiation from the Sun measured in watts per square metre at the planet's distance from the Sun. This is **not** the same as the amount of radiation hitting the planet's surface for two reasons. Firstly, some radiation is absorbed by the planet's atmosphere before it hits the surface. Secondly, the Sun is directly overhead only at tropical latitudes at midday. At other places the Sun's radiation is weaker because its rays are spread out over a larger region of the surface

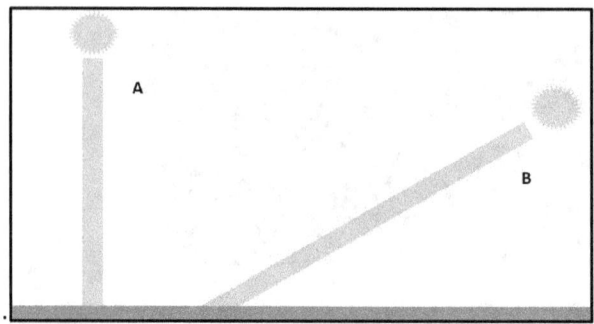

*At midday near the equator the Sun is directly overhead in the sky **(A)**. In the afternoon, as the Sun gets lower in the sky, its rays are spread out over a larger area **(B)**.*

4. The topographic range is the difference in altitude between the lowest and highest surface features. For example, on Earth this would be between the top of Mount Everest and the deepest point on the ocean floor.

Table II - Parameters relating to orbit and rotation

	Venus	Earth	Ratio (V/E)	Notes
Semi-major axis (10^6 km)	108.21	149.60	0.723	
Sidereal orbit period (days)	224.701	365.256	0.615	5
Tropical orbit period (days)	224.695	365.242	0.615	5
Perihelion (10^6 km)	107.48	147.09	0.731	
Aphelion (10^6 km)	108.94	152.10	0.716	
Synodic period (days)	583.92	n/a	n/a	6
Mean orbital velocity (km/s)	35.02	29.78	1.176	
Maximum orbital velocity (km/s)	35.26	30.29	1.164	
Minimum orbital velocity (km/s)	34.79	29.29	1.188	
Orbit inclination (degrees)	3.39	n/a	n/a	7
Orbit eccentricity	0.0067	0.0167	0.401	8
Rotation period (hrs)	-5832.6	23.9345	243.690	9
Length of day	116.8 days	1 day	116.8	10
Tilt of poles (degrees)	2.64	23.44	0.113	11

Notes

5. The sidereal orbital period or **sidereal year** is how long an object takes to complete one orbit around the Sun. The tropical orbital period or **tropical year** is the time that the Sun takes to return to the same position in the cycle of seasons; for example, the time from one spring equinox to the next spring equinox, or from one summer solstice to the next summer solstice. Because of a phenomenon known as the precession of the equinoxes, the tropical year is slightly shorter than the sidereal year.

6. The synodic period is the amount of time it takes Venus to gain a lap, with respect to Earth, in its orbit around the Sun.

7. This is the angle between the plane of Venus's orbit (shown in blue) and the plane of the Earth's orbit (shown in green).

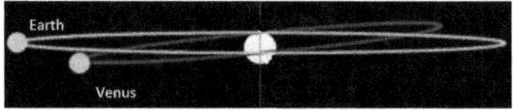

8. The eccentricity is a number between 0 and 1 and is a measure of how elliptical the planet's orbit is. If **a** is the length of short axis of the ellipse (known as the minor axis) and **b** is the length of the long axis of the ellipse (known as the major axis) then the eccentricity e is given by

$$e^2 = 1 - (a^2/b^2)$$

A circle has a=b and so could be thought of a special kind of ellipse with eccentricity 0.

9. The rotation period of Venus is a negative number because it rotates in the opposite direction to which it revolves around the Sun.

10. This is how long a day is on the planet.

11. This is how tilted the poles of the planet are with respect to the plane of its orbit.

Table III Atmospheric parameters

	Venus	Earth
Surface pressure	92 bars	1 bar
Mean surface temperature	464 degrees Celsius	15 degrees Celsius
Maximum surface temperature	464 degrees Celsius	57 degrees Celsius
Minimum surface temperature	464 degrees Celsius	-89 degrees Celsius
Constituents (by volume)	*Main constituents* carbon dioxide 96.5% nitrogen 3.5% *Trace gases* sulphur dioxide 150 ppm argon 70 ppm water vapour 20 ppm carbon monoxide 17 ppm helium 12 ppm neon 7ppm	*Main constituents* nitrogen 78.08 oxygen 20.95% argon 0.93% *Trace gases* carbon dioxide 412 ppm* Neon 18 ppm Helium 5 ppm Methane 2 ppm

Notes

The amount of water vapour in the Earth's atmosphere varies greatly. It is almost zero in the cold dry air near the South Pole in winter and can be as high as 7% in hot humid regions around the equator. Because of this variation, the contribution of water vapour has been ignored and the figures for the Earth are for dry air.

The figure for the amount of carbon dioxide in the atmosphere was the peak value in the year 2018 and was taken from the National Oceanic and Atmospheric Administration website https://www.esrl.noaa.gov/gmd/ccgg/trends/. The carbon dioxide concentration varies seasonally and reaches its maximum value in the early northern hemisphere spring. It is gradually increasing at a rate of around two ppm per year.

References

Atkinson N (2008) Colonizing Venus with floating cities. Available at: *http://www.universetoday.com/15570/colonizing-venus-with-floating-cities/* (Accessed: 01 February 2019)

Birch, P (1991) Terraforming Venus quickly. Available at: *http://www.orionsarm.com/fm_store/TerraformingVenusQuickly.pdf* (Accessed: 01 February 2019).

Davis, P (2018) *Magellan*. Available at: *https://solarsystem.nasa.gov/missions/magellan/in-depth/* (Accessed: 01 February 2019).

Drozdov A P, Eremets M I, Troyan I A, Ksenofontov V and Shylin S I (2015) Conventional superconductivity at 203 kelvin at high pressures in the sulfur hydride system. Available at:*http://www.nature.com/nature/journal/v525/n7567/full/nature14964.html* (Accessed: 01 February 2019)

Enerdata (2018) *Global energy statistical yearbook 2017 - Electricity domestic consumption,* Available at: *https://yearbook.enerdata.net/electricity-domestic-consumption-data-by-region.html* (Accessed: 01 February 2019).

Espenak, F (2014) 2018 calendar of astronomical events, Available at: *http://astropixels.com/ephemeris/astrocal/astrocal2018gmt.html* (Accessed: 01 February 2019).

Howell, E (2015) Venus' volcanoes are likely still active, Available at: *https://www.space.com/29742-venus-volcanoes-still-active.html* (Accessed: 01 February 2019).

Lorenz, R D (2018) *Lightning detection on Venus: a critical review,* Available at: https://progearthplanetsci.springeropen.com/articles/10.1186/s40645-018-0181-x (Accessed: 31 October 2018).

Luhmann, J G and Russell, C T (1997) *Venus: Magnetic Field and Magnetosphere,* Available at: http://www-ssc.igpp.ucla.edu/personnel/russell/papers/venus_mag/ (Accessed: 01 February 2019).

McClure, B (2012) Everything you need to know: Venus transit on June 5-6. Available at: http://earthsky.org/astronomy-essentials/last-transit-of-venus-in-21st-century-will-happen-in-june-2012 (Accessed: 01 February 2019).

NASA (2012) *Six millennium catalog of Venus transits: 2000 BCE to 4000 CE.* Available at: https://eclipse.gsfc.nasa.gov/transit/catalog/VenusCatalog.html (Accessed: 01 February 2019).

NASA (2013) *Ozone hole watch.* Available at: http://ozonewatch.gsfc.nasa.gov/facts/ozone_SH.html (Accessed: 01 February 2019)

U.S. Department of the Interior (2015) *How much water is there on, in, and above the Earth?,* Available at: http://water.usgs.gov/edu/earthhowmuch.html (Accessed: 01 February 2019)

Williams, D (2017) *Venus Fact Sheet.* Available at: http://nssdc.gsfc.nasa.gov/planetary/factsheet/venusfact.html (Accessed: 01 February 2019)

Williams, D R (2017) NASA space science data coordinated archive. Available at: https://nssdc.gsfc.nasa.gov/nmc/spacecraftDisplay.do?id=1962-041A (Accessed: 01 February 2019)

About the author

The author studied mathematics and astronomy at Warwick University, followed by a PhD in astronomy from Manchester University. He lives near Manchester in the North West of England.

The author writes a twice monthly popular science blog www.explainingscience.org. It is about astronomy, space travel and space exploration and is aimed at the non-scientist.

This is the author's third book. He has written two other books, *'The Moon'* and *'Is there anyone out there?'* These are available in Kindle format and can be downloaded from the Amazon Kindle bookstore.

www.ingramcontent.com/pod-product-compliance
Lightning Source LLC
Chambersburg PA
CBHW051324220526
45468CB00004B/1492